The Variational Theory of Geodesics

The Variational Theory of Geodesics

M.M. Postnikov
Professor, Moscow State University
Senior Associate, V. A. Steklov Institute of Mathematics
U.S.S.R. Academy of Sciences

Translated by Scripta Technica, Inc.
Edited by Bernard R. Gelbaum

Dover Publications, Inc., New York

Published in Canada by General Publishing Company, Ltd., 30 Lesmill Road, Don Mills, Toronto, Ontario.
Published in the United Kingdom by Constable and Company, Ltd., 10 Orange Street, London WC2H 7EG.

This Dover edition, first published in 1983, is an unabridged and unaltered republication of the work first published by W. B. Saunders Company, Philadelphia, in 1967 in the series Saunders Mathematics Books. (Original Russian: *Variatsionnaya Teoriya Geodezicheskikh*, Nauka Press, Moscow, 1965.)

Manufactured in the United States of America
Dover Publications, Inc., 180 Varick Street, New York, N.Y. 10014

Library of Congress Cataloging in Publication Data

Postnikov, M. M. (Mikhail Mikhaĭlovich)
The variational theory of geodesics.

Translation of: Variat͡sionnai͡a teorii͡a geodezicheskikh.
Reprint. Originally published: Philadelphia: Saunders, 1967.
1. Geodesics (Mathematics) 2. Calculus of variations. I. Gelbaum, Bernard R. II. Title.
QA649.P6713 1983 516.3′6 82-18215
ISBN 0-486-63166-4

FOREWORD

In this compact volume the author presents the most fundamental aspects of modern differential geometry as well as the basic tools required for the study of Morse theory.

The first half of the book contains an exposition of Riemannian geometry based on Koszul's axiom for an affine connection. The presentation is modeled after the treatment in S. Helgason's book, *Differential Geometry and Symmetric Spaces*, Academic Press, 1962.

The second half deals with Morse's variational theory of geodesics with significant amplifications given by Bott in his paper on the stable homotopy of the classical groups (Annals of Mathematics, 1959).

The presentation is self-contained and requires no prerequisites beyond a good course in calculus and some familiarity with point-set topology.

Bernard R. Gelbaum

CONTENTS

The Variational Theory of Geodesics

SMOOTH MANIFOLDS

1. Smooth Manifolds

Let M denote an arbitrary topological space and let f, f^1, ..., f^r denote real functions defined on M. We shall say that a function f *depends smoothly* on the functions f^1, ..., f^r if there exists an everywhere differentiable real function $u(t^1, \ldots, t^r)$ defined on R^r such that

$$f = u(f^1, \ldots, f^r) \text{ on } M,$$

that is, such that

$$f(p) = u(f^1(p), \ldots, f^r(p)) \tag{1}$$

for an arbitrary point $p \in M$. If equation (1) is valid only for points p in some open subset $U \subset M$, we shall say that the function f depends smoothly on the functions f^1, ..., f^r *on the set U*. If equation (1) is valid for points p in some neighborhood of a point $p_0 \in M$, we shall say that the function f depends smoothly on the functions f^1, ..., f^r close to the point p_0.

Analogously, we shall say that two functions *coincide close to a point* $p_0 \in M$ if they assume the same values in some neighborhood of that point.

We shall say that a Hausdorff space M is a *smooth premanifold* if the set of all real functions defined on the space M contains a nonempty subset $6 = 6(M)$ such that

(1) an arbitrary function that depends smoothly on the functions in 6 belongs to 6 and

(2) an arbitrary function that coincides, close to every point $p \in M$, with some function in 6 (generally speaking, depending on the point p) belongs to f.

We shall say that functions in 6 are smooth functions (on M). It follows from (1) that an arbitrary constant function belongs to 6 and that the operations of addition and multiplication of functions in 6 produce functions in 6. Consequently,

The set 6 is an algebra over the field R of real numbers.

This algebra is commutative and associative and it possesses a unit element (namely, the function that is identically equal to unity).

We note that the connection between smoothness and topology is rather weak [appearing only in condition (2)]. In particular, smooth functions are not necessarily continuous. A close correlation between smoothness and topology exists only in the case of manifolds (see below, section 2).

We obtain an important example of a smooth premanifold by considering an arbitrary m-dimensional linear space L. If we choose a basis in L, we can regard each function on L as a function of m real variables—the coordinates of the vectors in L relative to that basis. Obviously, the set $\mathfrak{S}(L)$ of all functions that are defined on L that are infinitely differentiable functions of these coordinates is independent of the basis chosen and it satisfies conditions (1) and (2) of the definition of a smooth premanifold. In what follows, when we speak of a linear space L as a smooth premanifold, we shall always assume that it is the functions in $\mathfrak{S}(L)$ that constitute the set of smooth functions on L.

In particular, the Euclidean space R^m, the points of which are the m-tuples $(t^1, \ldots, t^m)$ of real numbers is a smooth premanifold. The smooth functions on R^m are simply the infinitely differentiable functions of the variables $t^1, \ldots, t^m$.

Suppose now that K is an arbitrary subset of some smooth premanifold M. We shall say that a real function f defined on K is smooth (on K) if, close to an arbitrary point $p \in K$, it coincides with a smooth function on M, more precisely, if, for an arbitrary point $p \in K$, there exists in M a neighborhood U of p and a function g that is smooth on M such that $f(q) = g(q)$ for every point $q \in U \cap K$. It turns out that

The structure of a smooth premanifold is automatically set up in the set K (equipped with the topology induced from M).

To see this, suppose that the function f defined on K depends smoothly on functions $f^1, \ldots, f^r$ that are smooth on K; that is, suppose that

$$f = u(f^1, \ldots, f^r) \text{ on } K,$$

where u is some infinitely differentiable function of r real variables. By definition, for an arbitrary point $p \in K$ and arbitrary $i = 1, \ldots, r$, there exists in M a neighborhood U_i of the point p and a smooth function g^i on M such that

$$f^i|_{U_i \cap K} = g^i|_{U_i \cap K}.$$

Define

$$U = \bigcap_{i=1}^{r} U_i$$

and

$$g = u(g^1, \ldots, g^r).$$

Then, if $g \in \mathfrak{S}(M)$ and

$$f|_{U \cap K} = g|_{U \cap K}.$$

Thus, for an arbitrary point $p \in K$, there exists in M a neighborhood U of p and a smooth function g on M such that

$$f|_{U \cap K} = g|_{U \cap K}.$$

Consequently, the function f is a smooth function on K. Thus, we have shown that the set $\mathfrak{S}(K)$ of smooth functions of K satisfies condition (1) in the definition of a smooth premanifold.

Let us now verify condition (2). Suppose that the function f defined on K coincides close to an arbitrary point $p \in K$ with some function that is smooth on K; that is, suppose that, for an arbitrary point $p \in K$, there exists in K a neighborhood V of p and a smooth function h on K such that $f|_V = h|_V$. By definition, in M there is a neighborhood W of the point p and there exists a smooth function g on M such that $h|_{W \cap K} = g|_{W \cap K}$. Suppose that U is a neighborhood of the point p in M such that

$$U \cap K \subset W \cap V.$$

Then,

$$f|_{U \cap K} = g|_{U \cap K}.$$

Thus, close to an arbitrary point $p \in K$, the function f coincides with some function g that is smooth on M. Consequently, the function f is a smooth function on K. This completes the proof of the assertion made above.

Suppose that a function f is smooth in some neighborhood of a point $p \in M$ (and defined in that or some larger neighborhood). We shall say that such a function is smooth at the point p. In what follows, we shall find it convenient to treat two functions that are smooth at a point p and that coincide in some neighborhood of that point as the same function. Obviously, the set $\mathfrak{S}(p)$ of all such (equivalent classes of) functions also constitutes an algebra over the field R. Every function that is identically equal to one in some neighborhood of the point p is in the equivalence class of the unit element of that algebra.

Remark: In the terminology of the theory of sheaves, the algebra $\mathfrak{S}(p)$ is simply the algebra of germs of functions that are smooth in p, and the algebra $\mathfrak{S}(M)$ is the algebra of sections of the sheaf on M, the stalks of which are the algebras $\mathfrak{S}(p)$. We shall not use this terminology since we really do not need to introduce the material of the theory of sheaves.

Now, let N and M denote arbitrary premanifolds. We shall say that a continuous mapping Φ of the premanifold N into the premanifold M is smooth at a point $p \in N$ if, for an arbitrary function f that is smooth at the point $\Phi(p) \in M$, the composite function $f \circ \Phi$

is smooth at the point p. If a mapping $\Phi: N \to M$ is smooth at every point $p \in N$, we shall say that it is smooth on N. An example of a smooth mapping is the embedding mapping $\iota: K \to M$ into the premanifold M of an arbitrary subset $K \subset M$ equipped, in accordance with what has been said above, with the structure of a smooth premanifold. Another example is any smooth function f on M, treated as a mapping of the premanifold M into the real axis R.

We shall say that a homeomorphism $\Phi: N \to M$ of a premanifold N onto a premanifold M is a *diffeomorphism* if it is smooth on N and if its inverse $\Phi^{-1}: M \to N$ is smooth on M. We shall say that premanifolds N and M for which there exists at least one diffeomorphism $\Phi: N \to M$ are *diffeomorphic*.

2. Smooth Manifolds

A smooth premanifold M is called an *m-dimensional smooth manifold* if every point p in it possesses a neighborhood that is diffeomorphic to some open subset of the space R^m. We shall call any such neighborhood a *coordinate neighborhood* of the point p, and we shall call any diffeomorphism of it onto an open subset of the space R^m a *coordinate homeomorphism*. On every coordinate neighborhood U, a coordinate homeomorphism $\xi: U \to R^m$ defines by the formula

$$\xi(q) = (x^1(q), \ldots, x^m(q)), \quad q \in U, \tag{1}$$

m real functions $x^1 = x^1(q), \ldots, x^m = x^m(q)$. We shall call these functions the *local coordinates* (at the point p) corresponding to the coordinate homeomorphism ξ.

Let $x^1, \ldots, x^m$ denote arbitrary functions defined in some neighborhood U of a point $p \in M$. It is easy to see that

The functions $x^1, \ldots, x^m$ are local coordinates in the neighborhood U if and only if

(1) *The mapping $\xi: U \to R^m$ defined by formula (1) is a homeomorphism of the neighborhood U onto some open set $U^0 \subset R^m$,*

(2) *the functions $x^1, \ldots, x^m$ are smooth on U,*

(3) *each function f that is smooth at some point $q \in U$ depends smoothly on the functions $x^1, \ldots, x^m$ close to that point.*

Proof: If the functions $x^1, \ldots, x^m$ are local coordinates in the neighborhood U, then condition (1) is satisfied by definition, condition (2) is satisfied by virtue of the fact that $x^i = t^i \circ \xi$, where t^i is the function on R^m that assigns to each point in R^m its ith coordinate, and, finally condition (3) is satisfied by virtue of the fact that the smoothness of the inverse mapping $\xi^{-1}: U^0 \to U$ implies smoothness of the function $f \circ \xi^{-1}$ at the point $\xi(q) \in R^m$; that is, this function coincides close to that point with some smooth (i.e., infinitely differentiable) function u on R^m and, hence, $f = u \circ \xi$, that is, $f = u(x^1, \ldots, x^m)$ in the corresponding neighborhood of the point q.

Conversely, if the functions $x^1, \ldots, x^m$ satisfy conditions (1), (2) and (3), the mapping ξ is a diffeomorphism since condition (1) implies that it is a homeomorphism, condition (2) implies that it is smooth (since $u \circ \xi = u(x^1, \ldots, x^m)$ for a smooth function u defined on R^m), and condition (3) implies that the inverse mapping ξ^{-1} is also smooth (since the relation $f = u(x^1, \ldots, x^m)$ means that $u = f \circ \xi^{-1}$).

Obviously, at an arbitrary point p of a smooth manifold M, there exist infinitely many different systems of local coordinates. Suppose that we have a given local coordinate system $x^1, \ldots, x^m$ defined in a neighborhood U. To obtain a different system, we may, for example, take an arbitrary neighborhood V contained in the neighborhood U and consider the restrictions $y^1 = x^1|_V, \ldots, y^m = x^m|_V$ of the functions $x^1, \ldots, x^m$ to the neighborhood V. Obviously, the functions $y^1, \ldots, y^m$ are local coordinates in the neighborhood V. We may also take an arbitrary diffeomorphism φ of an open set $U^0 = \xi U$, where ξ is a coordinate homeomorphism defining local coordinates $x^1, \ldots, x^m$, onto some other open subset of the space R^m and consider the functions $z^1, \ldots, z^m$ defined by the diffeomorphism $\varphi \circ \xi$. These functions are local coordinates in the same neighborhood U as the coordinates $x^1, \ldots, x^m$. It is easy to see that these two transformations are sufficient to connect two arbitrary systems of local coordinates at the point p with each other. Specifically, if one system is defined in the neighborhood U by a coordinate homeomorphism ξ and the other is defined in a neighborhood V by a coordinate homeomorphism η, then the equation

$$\eta = \varphi \circ \xi,$$

where φ is the diffeomorphism

$$\eta \circ \xi^{-1}: \xi(U \cap V) \to \eta(U \cap V)$$

will hold throughout the neighborhood $U \cap V$.

These remarks, together with the familiar implicit-function theorem in analysis leads directly, in particular, to the following

Theorem (change of local coordinates). *Let $x^1, \ldots, x^m$ denote the local coordinates of a point p. Suppose that the functions*

$$y^i = u^i(x^1, \ldots, x^m), \quad i = 1, \ldots, m,$$

are smooth at p and possess the property that the Jacobian of the functions $u^1, \ldots, u^m$ is nonzero at the point $\xi(p) \in R^m$. Then, in some neighborhood of the point p, the functions $y^1, \ldots, y^m$ constitute a local coordinate system.

An example of local coordinates is provided by functions defined on some linear space L and assigning to each vector in that space its coordinates relative to a given basis. (Here, the corresponding coordinate neighborhood U coincides with the entire space L.) This shows that

An arbitrary m-dimensional linear space is an m-dimensional smooth manifold.

In particular, the Euclidean space R^m is an m-dimensional smooth manifold.

We note in conclusion that, obviously, local coordinates $x^1, \ldots, x^m$ on an arbitrary smooth manifold M are continuous in the corresponding coordinate neighborhood U.

From this it follows immediately that

An arbitrary smooth function f on a manifold M is continuous on M.

3. Open submanifolds. Property E

As we know, the structure of a smooth premanifold (cf. section 1) is introduced in a natural manner into an arbitrary subset of any premanifold M and hence of any smooth manifold. However, this premanifold will not, in general, be a smooth manifold even if the premanifold M is a smooth manifold. Nonetheless,

An arbitrary open subset W of a smooth m-dimensional manifold M is an m-dimensional smooth manifold.

Proof: For an arbitrary point $p \in W$, the intersection $U \cap W$, where U is an arbitrary coordinate neighborhood of the point p in the manifold M is obviously a coordinate neighborhood of the point p in the premanifold W. Here, the restrictions

$$x^1 |_{U \cap W}, \ldots, x^m |_{U \cap W}$$

of the local coordinates $x^1, \ldots, x^m$ defined in the neighborhood U will be local coordinates in the neighborhood $U \cap W$.

We shall say that smooth manifolds corresponding to open subsets W of a smooth manifold M are open submanifolds of the manifold M.

In accordance with the definition, the restriction $f|_W$ to an open submanifold W of an arbitrary smooth function on M is a smooth function on W. Consequently, the mapping $f \rightarrow f|_W$ defines a mapping (called the *restriction mapping*)

$$6(M) \rightarrow 6(W). \tag{1}$$

Obviously,

The mapping (1) is a homomorphism of the algebra $6(M)$ into the algebra $6(W)$.

In general, the mapping (1) is not an epimorphism; that is, not every function that is smooth on W is the restriction of some function that is smooth on M. We can assert only that it is locally epimorphic in the sense that

For every open set $V \subset W$ with compact[1] closure $\bar{V}$ also contained in W and for every function g that is smooth on W, there exists a smooth function f on M such that

$$f|_V = g|_V.$$

---

[1] A space or set is compact if every open covering admits a finite subcovering.

To prove this important assertion, we need the following

Lemma. *For an arbitrary compact subset C of a smooth manifold M and an arbitrary open set* $U \supset C$*, there exists a smooth function h on M that assumes values in the interval* $[0, 1] \subset R$ *and that is equal to one on C and equal to zero outside U.*

We shall first prove this lemma for the special case in which $M = R^m$. Suppose that $0 < a < b$ and that

$$F(t) = k \int_a^t \exp\left(-\frac{1}{(t-a)(t-b)}\right) dt, \quad a \leqslant t \leqslant b,$$

where the normalizing factor k is chosen in such a way that $F(b) = 1$. It is easy to see that all the derivatives of the function $F(t)$ vanish at the points a and b. Therefore, if we extend the definition of the function $F(t)$ for all values of $t \in R$ according to the formula

$$F(t) = \begin{cases} 0, & \text{if } t \leqslant a, \\ 1, & \text{if } t \geqslant b, \end{cases}$$

we obtain a function that is infinitely differentiable on the entire real axis, that assumes values only in the interval $[0, 1]$, that vanishes for $t \leqslant a$, and that is equal to one for $t \geqslant b$.

Now, let S and S' denote two concentric spheres in the space R^m. Let us construct the function $F(t)$ described above for the numbers a and b equal to the radii of the spheres S and S'. Let us now define a function g on R^m by

$$g(p) = F(r),$$

where r is the distance from the point p to the common center of the spheres S and S'. The function g is infinitely differentiable, it assumes values only in the interval $[0, 1]$, it vanishes inside the sphere S, and it is equal to one outside the sphere S'.

Consider now all open balls in the space R^m that contain at least one point of the compact set C and that have closures entirely contained in the open set U. Since these balls cover the set C, it follows, by virtue of the compactness C, that there is a finite system of such balls covering C. Let $S_1, \ldots, S_k$ denote the spheres constituting the boundaries of these balls. Since the set U is open, corresponding to each of the spheres S_i there exists a concentric sphere S_i' of larger radius such that the ball bounded by the sphere S_i' is, like the spheres S_i, contained in U. Let g_i denote the function g constructed for the spheres S_i and S_i'. Obviously, the function $g_1 g_2 \ldots g_k$ is infinitely differentiable, assumes values only in the interval $[0, 1]$, vanishes inside each sphere S_i and hence on the set C, and is equal to one outside all the spheres S_i' and, in particular, outside the set U. Consequently, the function $h = 1 - g_1 g_2 \ldots g_k$ possesses all the required properties.

Let us turn now to the general case of an arbitrary manifold M. Let p denote an arbitrary point in the set C and let U_p denote some coordinate neighborhood of it. Suppose further that V_p and W_p are neighborhoods of the point p such that $\overline{W}_p \subset V_p$ and $\overline{V}_p \subset U_p \cap U$ and that the closure $\overline{W}_p$ of the neighborhood W_p is compact. (Since the open set $U_p \cap U$ is homeomorphic to some open subset of the space R^m, such neighborhoods W_p and V_p exist.) Suppose, finally, that V_p^0 and W_p^0 are the images of the neighborhoods V_p and W_p under the coordinate homeomorphism $\xi: U_p \to R^m$. Since the set $\overline{W}_p^0 = \xi(\overline{W}_p)$ is compact and the set V_p^0 is open and contains the set $\overline{W}_p^0$, the special case, already proven, of the lemma is applicable to these sets. Therefore, there exists an infinitely differentiable function h_p on the space R^m that assumes values only in the interval $[0, 1]$, that is equal to one on the set $\overline{W}_p^0$, and that vanishes outside the V_p^0. Let us now define a real function g_p on the manifold M by setting

$$g_p(q) = \begin{cases} h_p(x^1(q), \ldots, x^m(q)), & \text{if} \quad q \in U_p, \\ 0, & \text{if} \quad q \notin U_p, \end{cases}$$

where $x^1, \ldots, x^m$ are local coordinates defined in the neighborhood U_p. Obviously, the function g_p is uniquely determined. Furthermore, close to an arbitrary point $q \in M$, it coincides either with the function $h_p(x^1, \ldots, x^m)$, which is smooth at q, or with the smooth function 0 (we recall that, by hypothesis, $\overline{V}_p \subset U_p$) and hence is a smooth function of M. It assumes values only in the interval $[0, 1]$, is equal to one on the set $\overline{W}_p$, and vanishes outside the set V_p.

Since the set C is compact, it contains a finite sequence of points $p_1, \ldots, p_k$ such that the sets $\overline{W}_{p_1}, \ldots, \overline{W}_{p_k}$ cover the entire set C. Let $V_{p_1}, \ldots, V_{p_k}$ denote the neighborhoods of V_p and let $g_{p_1}, \ldots, g_{p_k}$ denote the functions g_p corresponding to the points $p_1, \ldots, p_k$. Obviously, the function

$$h = 1 - (1 - g_{p_1}) \ldots (1 - g_{p_k})$$

assumes values only in the interval $[0, 1]$, is equal to one on the set $\overline{W}_{p_i}$ and hence on the set C, and vanishes outside the union of the sets V_{p_i}, in particular, outside the set U (since $V_{p_i} \subset U$). This completes the proof of the lemma.

It follows easily from the lemma that

An arbitrary smooth function g on a compact set $C \subset M$ is the restriction of some smooth function f on M.

Proof: For an arbitrary point $p \in C$, there exists in accordance with the definition of functions that are smooth on C a neighborhood $U_p \subset M$ and a smooth function g_p on M such that

$$g|_{U_p \cap C} = g_p|_{U_p \cap C}.$$

Let V_p be a neighborhood of the point p the closure $\overline{V}_p$ of which is compact and contained in U_p. According to the lemma we have just proved, there exists a smooth function h_p defined on the manifold M and assuming values only in the interval $[0, 1]$, equal to one on $\bar{V}_p$, and vanishing outside U_p. Since the set C is compact, there exists a finite sequence of points $p_1, \ldots, p_n \in C$ such that the set

$$V = \bigcup_{i=1}^{n} V_{p_i}$$

contains the set C. Consider the function

$$\tilde{g} = \frac{h_{p_1} g_{p_1} + \ldots + h_{p_n} g_{p_n}}{h_{p_1} + \ldots + h_{p_n}}.$$

Clearly, the denominator $h_{p_1} + \ldots + h_{p_n}$ is nonzero on V, so that this function is defined and smooth on V. Furthermore, for an arbitrary point $p \in C$ and arbitrary $i = 1, \ldots, n$, either $g_{p_i}(p) = g(p)$ (in case $p \in U_{p_i} \cap C$) or $h_{p_i}(p) = 0$ (in case $p \notin U_{p_i} \cap C$). Therefore, $\tilde{g}(p) = g(p)$; that is,

$$\tilde{g}|_C = g.$$

Let us now consider an arbitrary neighborhood W of the set C, the closure $\bar{W}$ of which is contained in the neighborhood V. Since the closure $\bar{W}$ is obviously compact, it follows from the lemma just proved that there exists a function h that is smooth on M, that is equal to unity on $\bar{W}$, and that vanishes outside V. Let us define a function f on M by

$$f(p) = \begin{cases} \tilde{g}(p)\, h(p), & \text{if} \quad p \in V, \\ 0, & \text{if} \quad p \notin V. \end{cases}$$

Obviously, close to arbitrary $p \in M$, this function coincides with some smooth function defined on M. Therefore, it is a smooth function of M itself. Furthermore,

$$f|_C = g.$$

That the restriction mapping (1) is a local epimorphism, as defined above, follows easily from the assertion just proved (one need only take for the set C the closure $\bar{V}$ of the neighborhood V).

Remark: That the mapping (1) is a local epimorphism can also be proved directly by repeating the second part of the above reasoning. The first part (the construction of the function $\tilde{g}$) is not necessary in this special case.

We shall say that a smooth manifold N has property E (the letter E standing for "extension") if, for an arbitrary smooth manifold M and an arbitrary compact subset $C \subset M$, every smooth mapping $\Psi: C \to N$ is the restriction of some smooth mapping

$\Phi: M \to N$. In this terminology, the assertion proven above means that

The real line R has property E.

Furthermore, it is easy to see that

For an arbitrary natural number n, the Euclidean n-dimensional space R^n has property E.

Proof: Any smooth mapping $C \to R^n$ is defined by n smooth functions on C. To extend such a mapping to the entire manifold M, we need only extend these functions to M.

One can also show without difficulty that

The open unit ball E^n in the space R^n (that is, the set of all points $(t^1, \ldots, t^n) \in R^n$ such that $(t^1)^2 + \ldots + (t^n)^2 < 1$) has property E.

Proof: Obviously, an arbitrary manifold that is diffeomorphic to a manifold possessing property E has property E itself. On the other hand, if we assign to each point $(t^1, \ldots, t^n) \in R^n$ the point $(\tau^1, \ldots, \tau^n) \in E^n$ with coordinates

$$\tau^i = \frac{2}{\pi} \frac{\arctan|t|}{|t|} t^i, \quad i = 1, \ldots, n,$$

where $|t| = \sqrt{(t^1)^2 + \ldots + (t^n)^2}$, we obviously obtain a diffeomorphism from R^n onto E^n.

Let us show now that, in general,

Every open convex subset D of the space R^n has property E.

Proof: Let us suppose first that the set D is bounded. We shall show that it is diffeomorphic to the ball E^n. From the elementary theory of convex bodies, we know that the set D is starlike with respect to any interior point p_0 belonging to it; that is, an arbitrary ray issuing from the point p_0 intersects the boundary of D at exactly one point. Each such ray L_p is determined by some point p on the unit sphere S^{n-1}. Let $\varphi(p)$ denote the distance from the point p_0 to the point at which the ray L_p intersects the boundary of the set D. Simple geometric reasoning shows that the function $\varphi(p)$ is continuous on the sphere S^{n-1} and in fact satisfies a Lipschitz condition; that is, there exists a number L such that

$$|\varphi(p) - \varphi(q)| < L\,|p - q|,$$

where $|p - q|$ is the Euclidean distance between the points p and q in S^{n-1}. On the other hand, we can show that

For an arbitrary function φ defined on the sphere S^{n-1} and satisfying a Lipschitz condition, there exists a continuous function g defined in the closed unit ball $\bar{E}^n$ such that

(1) *the function g is smooth in the open ball E^n;*

(2) *the function g coincides with the function φ on the sphere S^{n-1};*

(3) *the function g is nonzero at the point* $(0, \ldots, 0)$;

(4) *the function g increases monotonically along each radius of the ball $\bar{E}^n$.*

Since an arbitrary point in the ball $\bar{E}^n$ can be defined by its distance r from the coordinate origin $(0, \ldots, 0)$ and some point $p \in S^{n-1}$, we can define a function g satisfying these conditions by

$$g(r, p) = A\left(e^{1-\frac{1}{r^2}} - 1\right) + \\ + \frac{2^n e^{1-\frac{1}{r^2}}}{(1-r)^n} \int_{-\infty}^{+\infty} \cdots \int_{-\infty}^{+\infty} \left[\prod_{i=1}^{n} \omega\left(\frac{2(t^i - \xi^i)}{1-r}\right)\right] \varphi(\tilde{t})\, dt^1 \ldots dt^n,$$

where

(a) A is some sufficiently large positive number;

(b) $\xi^1, \ldots, \xi^n$ are the coordinates of the point $p \in S^{n-1}$;

(c) $\tilde{t}$ is the point on the sphere S^{n-1} with coordinates $\frac{t^1}{|t|}, \ldots, \frac{t^n}{|t|}$, where

$$|t| = \sqrt{(t^1)^2 + \ldots + (t^n)^2};$$

(d) $\omega(\tau)$ is a smooth function defined by

$$\omega(\tau) = \begin{cases} \frac{1}{k} e^{\frac{1}{\tau^2 - 1}}, & \text{if} \quad |\tau| \leqslant 1, \\ 0, & \text{if} \quad |\tau| \geqslant 1, \end{cases} \qquad k = \int_{-1}^{+1} e^{\frac{1}{\tau^2-1}}\, d\tau.$$

We omit the proof (quite elementary) that the function $g(r, p)$ thus defined does satisfy conditions (1)–(4). We note only that the verification rests primarily on the assumption that φ satisfies a Lipschitz condition.

Let us construct the function g for the function φ (corresponding to the convex set D). Then, let us define a mapping

$$\Phi : E^n \to D,$$

by taking for the image under Φ of an arbitrary point $(r, p) \in E^n$ that point on the ray L_p lying at a distance $g(r, p)$ from the point p_0. One can show without difficulty that this mapping is a diffeomorphism. Thus, the set D is diffeomorphic to the sphere E^n and hence has property E.

Let us suppose now that the set D is unbounded. Consider an arbitrary smooth manifold M, and arbitrary compact subset C of it, and an arbitrary smooth mapping $\Psi : C \to D$. Since the set C is compact and the mapping Ψ is continuous, the image $\Psi(C)$ of the set C under the mapping Ψ is compact and hence bounded; that is, it is contained in some open ball in the space R^n of sufficiently great radius. Let D' denote the intersection of the set D and this ball. Since $\Psi(C) \subset D'$, we can consider the mapping Ψ as a mapping from C into D'. However, since the set D' is open, convex, and bounded, it has property E by virtue of what we have proved above. Therefore, there exists a smooth mapping $\Phi : M \to D'$ such that $\Phi|_C = \Psi$. It remains only to point out that, since $D' \subset D$, we can assume that the mapping Φ is a mapping into the set D. This completes the proof of the assertion made above.

Remark: It can be shown that an unbounded convex set D is also diffeomorphic to the ball E^n. Since we shall not need this fact, we state it without proof.

4. Vector fields

Let M denote an arbitrary smooth manifold. We define a vector field on the manifold M as any derivation of the algebra $6(M)$, that is, a linear mapping X of this algebra into itself such that

$$X(fg) = f \cdot Xg + Xf \cdot g$$

for arbitrary functions f and g in $6(M)$. Obviously, the set $6^1(M)$ of all vector fields on the manifold M is an $6(M)$-module with respect to the operations

$$(X+Y)g = Xg + Yg, \qquad X, Y \in 6^1(M), \qquad g \in 6(M),$$
$$(fX)g = f \cdot Xg, \qquad\qquad X \in 6^1(M), \quad f, g \in 6(M).$$

Furthermore, one can easily see that, for arbitrary vector fields X and Y in $6^1(M)$, the mapping $[X, Y] : 6(M) \to 6(M)$ defined by

$$[X, Y]g = X(Yg) - Y(Xg), \quad g \in 6(M),$$

is also a vector field and that the operation $[X, Y]$ introduces into the module $6^1(M)$ the structure of a Lie algebra; that is, this operation is anticommutative:

$$[X, Y] = -[Y, X], \quad X, Y \in 6^1(M),$$

and satisfies Jacobi's identity

$$[[X, Y], Z] + [[Y, Z], X] + [[Z, X], Y] = 0, \quad X, Y, Z \in 6^1(M).$$

Let W be an arbitrary open submanifold of the manifold M and let X be some vector field on the manifold M. Let g be any smooth function on W. In accordance with section 3, for every point $p \in W$ there exist smooth functions f defined on the manifold M, such that $f|_V = g|_V$, where V is some neighborhood of the point p with compact closure $\overline{V} \subset W$. Let f_1 and f_2 denote smooth functions on M with this property. According to the lemma in section 3, for an arbitrary point $q \in V$, there exists a smooth function h on the manifold M that is equal to one at the point q and equal to 0 outside the neighborhood V. Then, $h(f_1 - f_2) = 0$ on M and, therefore,

$$h \cdot X(f_1 - f_2) + Xh \cdot (f_1 - f_2) = 0 \quad \text{on} \quad M.$$

From the facts that $h(q) = 1$, $f_1(q) = f_2(q)$, and $X(f_1 - f_2) = Xf_1 - Xf_2$, it follows that

$$(Xf_1)(q) - (Xf_2)(q) = (Xh)(q) \cdot (f_2(q) - f_1(q)) = 0,$$

that is,

$$(Xf_1)(q) = (Xf_2)(q).$$

Therefore, if we set

$$\bar{g}(q) = (Xf)(q), \quad q \in V,$$

where f is an arbitrary smooth function on M such that $f|_V = g|_V$, we obtain an unambiguously defined smooth function $\bar{g}$ on V. Obviously, the functions $\bar{g}$ corresponding to different neighborhoods V of the point p assume the same value at the point p. Therefore, if we set,

$$(Yg)(p) = \bar{g}(p), \quad p \in V,$$

we unambiguously define a function Yg on W. Since this function coincides close to an arbitrary point $p \in W$ with the function $\bar{g}$, which is smooth at p, the function Yg is smooth on W. Therefore, the mapping $g \to Yg$ defines a mapping

$$Y : 6(W) \to 6(W).$$

It is easily verified that this mapping is a vector field on the manifold W.

We shall call the field Y a *restriction* of the field X to the open submanifold W and we shall denote it by $X|_W$ and even simply by X if there is no danger of misunderstanding. One can easily verify that the restriction mapping

$$6^1(M) \to 6^1(W), \tag{1}$$

that assigns to each field $X \in 6^1(M)$ the field $X|_W \in 6^1(W)$ conserves all the algebraic operations listed above; that is, it is both a homomorphism of the modules (if the $6(W)$-module $6^1(W)$ is considered, on the basis of the restriction mapping $6(M) \to 6(W)$, as an $6(M)$-module) and a homomorphism of Lie algebras.

Just as with functions, the mapping (1) is not in general epimorphic; that is, not every field $Y \in 6^1(W)$ is the restriction of some field $X \in 6^1(M)$. The mapping (1) is, in general, only locally epimorphic; that is,

For any open set $\bar{V}$ whose closure V is compact and contained in W and for any field $Y \in 6^1(W)$, there exists a field $X \in 6^1(M)$ such that

$$X|_V = Y|_V.$$

Proof: According to the lemma in section 3, there exists a smooth function h defined on the manifold M that is equal to unity on $\bar{V}$, that vanishes outside some open set $U \supset \bar{V}$ the closure $\bar{U}$ of which belongs to W. It is easy to see that the formula

$$(Xf)(p) = \begin{cases} h(p)(Y(f|_W))(p), & \text{if } p \in W, \\ 0, & \text{if } p \notin W, \end{cases}$$

where f is an arbitrary smooth function on M, defines a vector field X on M with the required property.

5. Vector fields on coordinate neighborhoods

In the particular case in which an open submanifold W is a coordinate neighborhood U, it is easy to describe all the vector fields on W. Suppose that $x^1, \ldots, x^m$ are local coordinates defined in a neighborhood U. Then, for an arbitrary function f that is smooth on U and an arbitrary point $p \in U$, there exists a neighborhood $V \subset U$ of the point p such that throughout V the function f depends smoothly on the coordinates $x^1, \ldots, x^m$, that is, such that

$$f = u(x^1, \ldots, x^m) \text{ on } V,$$

where $u = u(t', \ldots, t^m)$ is a smooth function (uniquely defined by the function f) on R^m. Let V_1 denote another neighborhood of the point p with the same property; that is, $f = u_1(x^1, \ldots, x^m)$ on V_1, where $u_1 = u_1(t^1, \ldots, t^m)$ is some smooth function on R^m. Then, it is clear that the functions u and u_1 coincide in some neighborhood of the point $\xi(p) = (x^1(p), \ldots, x^m(p))$ [specifically, in the neighborhood $\xi(V \cap V_1)$, which is the image under the coordinate homeomorphism $\xi: U \to R^m$ of the neighborhood $V \cap V_1$ of the point p]. Therefore, their partial derivatives $\partial u / \partial t^i$ and $\partial u_1 / \partial t^i$, for arbitrary $i = 1, \ldots, m$, assume the same value at the point $\xi(p)$. It follows from this that the formula

$$h_i(p) = \left(\frac{\partial u}{\partial t^i} \right)_{\xi(p)}, \quad i = 1, \ldots, m$$

defines functions $h_1, \ldots, h_m$ unambiguously on U. In the neighborhood V, the functions h_i coincide with the smooth functions $\partial u(x^1, \ldots, x^m) / \partial t^i$ and hence are smooth functions at p. Since the point p is arbitrary, this means that each function h_i is smooth on U. We shall denote this function by $\partial f / \partial x^i$ and call it the partial derivative of the function f with respect to the coordinate x^i. In accordance with this definition,

If

$$f = u(x^1, \ldots, x^m) \quad \textit{on} \quad V \subset U,$$

then

$$\frac{\partial f}{\partial x^i} = \frac{\partial u}{\partial x^i}(x^1, \ldots, x^m) \quad \textit{on} \quad V.$$

One can easily verify that the mapping

$$\frac{\partial}{\partial x^i} : f \to \frac{\partial f}{\partial x^i}, \qquad i = 1, \ldots, m$$

is a vector field on U. The m vector fields

$$\frac{\partial}{\partial x^1}, \ldots, \frac{\partial}{\partial x^m}$$

that we have constructed obviously satisfy the relation

$$\frac{\partial x^j}{\partial x^i} = \delta_i^j, \qquad i, j = 1, \ldots, m,$$

where δ_i^j is the Kronecker delta:

$$\delta_i^j = \begin{cases} 0 \text{ for } i \neq j \\ 1 \text{ for } i = j \end{cases}$$

From this it follows immediately that

The vector fields $\frac{\partial}{\partial x^1}, \ldots, \frac{\partial}{\partial x^m}$ *are linearly independent* (in the $\mathfrak{S}(U)$-module $\mathfrak{S}^1(U)$).

Remark: In the particular case in which $M = R^m$, we may take for the neighborhood U the entire space R^m and we may take for local coordinates $x^1, \ldots, x^m$ the functions that map each point $(t^1, \ldots, t^m) \in R^m$ into its coordinates $t^1, \ldots, t^m$. Obviously, in this case,

The operations $\frac{\partial}{\partial x^1}, \ldots, \frac{\partial}{\partial x^m}$ *coincide with the usual operations of partial differentiation with respect to variables* $t^1, \ldots, t^m$.

In particular, for $m = 1$, there is only one coordinate $t = t^1$, and we obtain

The operation $\frac{\partial}{\partial t}$ *is simply differentiation* $\frac{d}{dt}$ *with respect to the variable t.*

Let us return to a smooth function f defined on U and to a neighborhood V in which $f = u(x^1, \ldots, x^m)$. We may assume without loss of generality that the image ξV of the neighborhood V under the coordinate homeomorphism

$$\xi : q \to (x^1(q), \ldots, x^m(q))$$

is an open ball in the space R^m with center at the point $\xi(p) = (a^1, \ldots, a^m)$. In this open ball, the function $u(t^1, \ldots, t^m)$ admits the representation[1]

---

[1] We are adopting the summation convention of tensor analysis; that is, when a single letter is used once above and once below, summation with respect to that letter is understood. For example, $(x^i - a^i)\int_0^1 \frac{\partial u}{\partial t^i} (\ldots)$ means $\sum_{i=1}^m (t^i - a^i) \int_0^i \frac{\partial u}{\partial t^i}$.

$$u(t^1, \ldots, t^m) = u(a^1, \ldots, a^m) +$$
$$+ \int_0^1 \frac{\partial}{\partial s} u(a^1 + s(t^1 - a^1), \ldots, a^m + s(t^m - a^m))\, ds =$$
$$= u(a^1, \ldots, a^m) +$$
$$+ (t^i - a^i) \int_0^1 \frac{\partial u}{\partial t^i}(a^1 + s(t^1 - a^1), \ldots, a^m + s(t^m - a^m))\, ds.$$

Therefore, when we have defined on V the smooth functions $g_i = g_i(q)$ by

$$g_i(q) = \int_0^1 \frac{\partial u}{\partial t^i}(a^1 + s(x^1(q) - a^1), \ldots$$
$$\ldots, a^m + s(x^m(q) - a^m))\, ds, \qquad i = 1, \ldots, m,$$

we obtain

$$f = f(p) + (x^i - a^i) g_i \quad \text{on } V, \tag{1}$$

where the numbers $f(p)$ and a^i are considered as constant functions (assuming the values $f(p)$ and a^i, respectively) on V.

Now, let X denote an arbitrary vector field on U. Obviously, the formula

$$X_p f = (Xf)(p), \quad p \in U, \quad f \in \mathfrak{G}(p),$$

where the X on the right denotes the restriction of the field X to the corresponding neighborhood of the point p, defines a linear mapping X_p of the algebra $\mathfrak{G}(p)$ into the field R such that

$$X_p(fg) = X_p f \cdot g(p) + f(p) \cdot X_p g \tag{2}$$

for arbitrary functions $f, g \in \mathfrak{G}(p)$.

From equation (2) it follows immediately that

$X_p(c) = 0$ *for every constant c.*

Therefore, if we apply the mapping X_p to both sides of formulas (1), we obtain

$$X_p f = X_p x^i \cdot g_i(p).$$

But

$$g_i(p) = \int_0^1 \frac{\partial u}{\partial t^i}(a^1, \ldots, a^m)\, ds = \frac{\partial u}{\partial t^i}(a^1, \ldots, a^m) = \frac{\partial f}{\partial x^i}(p).$$

Thus,

$$X_p f = X_p x^i \cdot \frac{\partial f}{\partial x^i}(p). \tag{3}$$

If f in this formula denotes any function that is smooth on U, we immediately obtain, by virtue of the definition of the symbol X_p,

$$X = Xx^i \cdot \frac{\partial}{\partial x^i} \quad \text{on} \quad U. \tag{4}$$

Since the fields $\frac{\partial}{\partial x^1}, \ldots, \frac{\partial}{\partial x^m}$ are linearly independent, this means that

The vector fields $\frac{\partial}{\partial x^1}, \ldots, \frac{\partial}{\partial x^m}$ *constitute a basis for the module* $\mathfrak{G}^1(U)$ *over the algebra* $\mathfrak{G}(U)$; *an arbitrary field* $X \in \mathfrak{G}^1(U)$ *is expressed in terms of this basis in accordance with formula (4).*

In the particular case in which $M = R^m$, we have

The partial differentiations $\frac{\partial}{\partial t^1}, \ldots, \frac{\partial}{\partial t^m}$ *constitute a basis for the module* $\mathfrak{G}^1(R^m)$.

A basis $X_1, \ldots, X_m$ for the module $\mathfrak{G}^1(U)$ is said to be ***holonomic*** if there exists in U a system of local coordinates $x^1, \ldots, x^m$ such that

$$X_1 = \frac{\partial}{\partial x^1}, \ldots, X_m = \frac{\partial}{\partial x^m};$$

Otherwise, the basis $X_1, \ldots, X_m$ is said to be ***nonholonomic***.

In an arbitrary basis $X_1, \ldots, X_m$ of the module $\mathfrak{G}^1(U)$, every field $X \in \mathfrak{G}^1(U)$ is of the form

$$X = X^i X_i,$$

where the X^i are smooth functions on U. These functions are uniquely determined by the field X and are called the ***components*** of the field X relative to the basis $X_1, \ldots, X_m$. From formula (4),

$$X^i = Xx^i$$

for a holonomic basis.

The linear operations defined in the module $\mathfrak{G}^1(U)$ are expressed in terms of the components according to the formulas

$$(fX)^i = fX^i, \qquad f \in \mathfrak{G}(U), \quad X \in \mathfrak{G}^1(U),$$
$$(X+Y)^i = X^i + Y^i, \qquad X, Y \in \mathfrak{G}^1(U).$$

If a basis $X_1, \ldots, X_m$ *is holonomic, then*

$$[X, Y]^i = X^j \frac{\partial Y^i}{\partial x^j} - Y^j \frac{\partial X^i}{\partial x^j} \tag{5}$$

where $[X, Y]$ *denotes a Lie operation.*

This is true because

$$[X, Y]^i = [X, Y]\, x^i = X(Yx^i) - Y(Xx^i) = X(Y^i) - Y(X^i).$$

In particular, it follows from formula (5) that
For an arbitrary holonomic basis $X_1, \ldots, X_m$,

$$[X_i, X_j]=0. \tag{6}$$

In the general case,

$$[X_i, X_j]=c^k_{ij}X_k,$$

where the c^k_{ij} are smooth functions on U and formula (5) is replaced by the formula

$$[X, Y]^i=X^jX_j(Y^i)-Y^jX_j(X^i)+2c^i_{jk}X^jY^k.$$

Remark: Formula (6) means that the result of applying the two operations $\frac{\partial}{\partial x^i}$ and $\frac{\partial}{\partial x^j}$ to an arbitrary function g that is smooth on U is independent of the order in which these operations are performed. In other words, just as in elementary calculus,

$$\frac{\partial^2 g}{\partial x^i\,\partial x^j}=\frac{\partial^2 g}{\partial x^j\,\partial x^i}.$$

6. Vectors

We shall call the linear mapping introduced above

$$X_p : \mathfrak{S}(p)\to R$$

the vector of the field X at the point p. We shall call the linear mapping $A : \mathfrak{S}(p)\to R$ a *vector of the manifold* M at the point p if, in some neighborhood of the point p (or, equivalently, on the entire manifold M), there exists a vector field X such that $A=X_p$. It turns out that

A linear mapping $A : \mathfrak{S}(p)\to R$ *is a vector of the manifold M at a point p if and only if*

$$A(fg)=Af\cdot g(p)+f(p)\cdot Ag$$

for all f and g in $\mathfrak{S}(p)$.

The necessity of this condition follows immediately from relation (2) of section 5. On the other hand, in proving formula (3) of section 5, we used only relation (2), so that, for an arbitrary linear mapping A that satisfies the conditions of the statement above, we have the analogous formula

$$Af=a^i\frac{\partial f}{\partial x^i}(p), \text{where } a^i=Ax^i, \tag{1}$$

Therefore, the vector field X defined on a coordinate neighborhood U by the formula

$$X = a^i \frac{\partial}{\partial x^i},$$

has the property that $X_p = A$.

The set M_p of all vectors of the manifold M at the point p is obviously a linear space over the field R. This space is called the ***tangent space*** to the manifold M at the point p. In particular, it contains the vectors

$$\left(\frac{\partial}{\partial x^1}\right)_p, \ldots, \left(\frac{\partial}{\partial x^m}\right)_p, \tag{2}$$

where $x^1, \ldots, x^m$ is an arbitrary system of local coordinates at the point p. The vectors (2) are linearly independent, and any vector A in$\in M_p$ can be expressed as a linear combination of them:

$$A = Ax^i \cdot \left(\frac{\partial}{\partial x^i}\right)_p$$

[cf. formula (1)]; that is, they constitute a basis for the space M_p. Thus we have proven

For an arbitrary point $p \in M$, the dimension of the space M_p is equal to the dimension m of the manifold M.

We shall also denote the space M_p by the symbol $\mathfrak{6}^1(p)$.

Suppose, in particular, that the manifold M is an m-dimensional linear space L.

It is easy to see that for an arbitrary vector $A \in L$ and an arbitrary function f that is smooth on L, the function $\hat{A}f$ defined by

$$(\hat{A}f)(B) = \left.\frac{df(B + tA)}{dt}\right|_{t=0}, \quad B \in L,$$

is smooth on L and that the mapping $\hat{A}: \mathfrak{6}(L) \to \mathfrak{6}(L)$ is a vector field on L. Let $A_1, \ldots, A_m$ denote an arbitrary basis in the space L. If we set

$$f(B) = f(b^1, \ldots, b^m),$$

where $b^1, \ldots, b^m$ are the coordinates of the vector B relative to the basis $A_1, \ldots, A_m$, we immediately obtain

$$\hat{A}_i f = \frac{\partial f}{\partial b^i}.$$

From this it follows that

For an arbitrary vector $B \in L$, the vectors $(\hat{A}_1)_B, \ldots, (\hat{A}_m)_B$ are linearly independent and they constitute a basis for the tangent space L_B.

This means in particular, that

The mapping $A \to (\hat{A}_B)$ of the space L into the space L_B is an isomorphism.

In what follows, we shall usually identify the vectors A and $(\hat{A})_B$. Thus, we shall assume that $L = L_B$.

Returning to an arbitrary smooth manifold M, we note that, for an arbitrary vector field $X \in \mathfrak{G}^1(M)$, the function $p \to X_p$ that assigns to every point $p \in M$ the vector $X_p \in M_p$ obviously satisfies the following "smoothness condition":

For every function f that is smooth on M, the function $g(p) = X_p(f)$ is also a smooth function on M.

Conversely, as one can easily see,

Every function $p \to X_p$ that assigns to an arbitrary point $p \in M$ some vector $X_p \in M_p$ and satisfies the smoothness condition stated above corresponds to some vector field $X \in \mathfrak{G}^1(M)$.

This field X is obviously given by the formula

$$(Xf)(p) = X_p(f), \quad f \in \mathfrak{G}^1(M), \quad p \in M.$$

Thus, we shall treat the fields $X \in \mathfrak{G}^1(M)$ as "smooth functions" $p \to X_p$.

The concept of a vector at a point enables us to give a simple but often convenient necessary and sufficient condition for a system of m vector fields $X_1, \ldots, X_m$ defined on a coordinate neighborhood U to constitute a basis for the module $\mathfrak{G}^1(U)$. Specifically, as one can easily see,

Vector fields $X_1, \ldots, X_m \in \mathfrak{G}^1(U)$ constitute a basis for the $\mathfrak{G}(U)$-module $\mathfrak{G}^1(U)$ if and only if, for each point $q \in U$, the vectors

$$(X_1)_q, \ldots, (X_m)_q$$

constitute a basis for the space M_q.

Proof: Suppose that the fields $X_1, \ldots, X_m$ constitute a basis for the $\mathfrak{G}(U)$-module $\mathfrak{G}^1(U)$. Let A denote an arbitrary vector in the space M_q. By definition, there exists on U a vector field X such that $X_q = A$. Suppose that

$$X = X^i X_i$$

is the decomposition of the field X relative to the basis $X_1, \ldots, X_m$. Then,

$$A = X^i(q)(X_i)_q.$$

Consequently, the vectors $(X_1)_q, \ldots, (X_m)_q$ generate the space M_q. Since there are m of them, they are linearly independent and thus constitute a basis.

Conversely, suppose that the fields $X_1, \ldots, X_m \in \mathfrak{G}^1(U)$ possess the property that, for an arbitrary point $q \in U$, the vectors $(X_1)_q, \ldots, (X_m)_q$ constitute a basis for the space M_q. Let X denote an arbitrary vector field on U. Then, at an arbitrary point $q \in U$, we have the expansion

$$X_q = \hat{X}^i(q)(X_i)_q,$$

where the $\hat{X}^i(q)$ are numbers depending on the point q. Suppose that $x^1, \ldots, x^m$ are local coordinates defined in a neighborhood U. Then, for arbitrary $j = 1, \ldots, m$,

$$X_q x^j = \hat{X}^i(q)(X_i)_q x^j.$$

But, as we know, the numbers $X_q x^j$ are the values at the point q of the components $X^j = Xx^j$ of the field X relative to the basis $\frac{\partial}{\partial x^1}, \ldots, \frac{\partial}{\partial x^m}$. Analogously, $(X_i)_q x^j = X_i^j(q)$, where the X_i^j are the components of the field X_i relative to the basis $\frac{\partial}{\partial x^1}, \ldots, \frac{\partial}{\partial x^m}$. Thus,

$$X^j(q) = \hat{X}^i(q) X_i^j(q). \tag{3}$$

Clearly, the determinant $\det|X_i^j(q)|$ at an arbitrary point $q \in U$ will be nonzero. Therefore, we may use the familiar rule of Cramer to express the numbers $\hat{X}^i(q)$ in terms of the numbers $X^j(q)$ and $X_i^j(q)$. Since the components X^j and X_i^j are smooth functions of the point q, this means that the functions $\hat{X}^i = \hat{X}^i(q)$ also depend smoothly on q. Thus,

$$X = \hat{X}^i X_i,$$

where the $\hat{X}^i$ are functions that are smooth on U. Consequently, the fields $X_1, \ldots, X_m$ constitute a basis for the $\mathfrak{G}(U)$-module $\mathfrak{G}^1(U)$.

We note also, by virtue of the arbitrariness (described in section 2) in the choice of local coordinates at the point p of the manifold M:

For an arbitrary basis $A_1, \ldots, A_m$ for the space M_p, there exists at the point p a system of local coordinates $x^1, \ldots, x^m$ such that

$$\left(\frac{\partial}{\partial x^1}\right)_p = A_1, \ldots, \left(\frac{\partial}{\partial x^m}\right)_p = A_m.$$

In other words, an arbitrary basis for the space M_p is, as we say, holonomic at the point p.

In conclusion, let us consider an arbitrary mapping Φ, smooth at the point $p \in N$, of a manifold N into a manifold M. To an arbitrary vector $A \in N_p$ we can assign the mapping

$$d\Phi_p(A) : \mathfrak{G}(\Phi(p)) \to \mathfrak{G}(\Phi(p)),$$

by setting

$$d\Phi_p(A)(g) = A(g \circ \Phi)$$

for an arbitrary function g that is smooth at the point $\Phi(p) \in M$. Obviously, $d\Phi_p(A)$ is a vector in the manifold M at the point $\Phi(p)$, and the mapping obtained

$$d\Phi_p : N_p \to M_{\Phi(p)}$$

of the space N_p into the space $M_{\Phi(p)}$ is linear. It is called the *differential* of the mapping Φ at the point p.

One can easily see that, for arbitrary smooth mapping $\Phi : N \to M$ and $\Psi : P \to N$ and for an arbitrary point $p \in P$,

$$d(\Phi \circ \Psi)_p = d\Phi_{\Psi(p)} \circ d\Psi_p.$$

7. Linear differential forms

As above, let M denote an arbitrary smooth manifold of dimension m.

We shall refer to an arbitrary $\mathfrak{S}(M)$-linear mapping of the $\mathfrak{S}(M)$-module $\mathfrak{S}^1(M)$ into the algebra $\mathfrak{S}(M)$ (considered as an $\mathfrak{S}(M)$-module) as a ***linear differential form*** on the manifold M. Thus, every form ω assigns to an arbitrary vector field $X \in \mathfrak{S}^1(M)$ some function $\omega(X)$ that is smooth on M. Hence, for an arbitrary function f that is smooth on M and arbitrary fields $X, X_1, X_2 \in \mathfrak{S}^1(M)$,

$$\omega(fX) = f\omega(X),$$
$$\omega(X_1 + X_2) = \omega(X_1) + \omega(X_2).$$

Obviously, the set $\mathfrak{S}_1(M)$ of all linear differential forms on the manifold M is an $\mathfrak{S}(M)$-module with respect to the operations

$$(\omega_1 + \omega_2)(X) = \omega_1(X) + \omega_2(X), \quad \omega_1, \omega_2 \in \mathfrak{S}_1(M), \quad X \in \mathfrak{S}^1(M),$$
$$(f\omega)(X) = f \cdot \omega(X), \quad \omega \in \mathfrak{S}_1(M), \quad f \in \mathfrak{S}(M), \quad X \in \mathfrak{S}^1(M).$$

Now, let W denote an arbitrary open submanifold of the manifold M and let ω denote any linear differential form on M. Let Y denote an arbitrary vector field on W and let p denote an arbitrary point in W. Consider a neighborhood V of the point p such that $\overline{V}$ is a compact subset of W. Then, as was shown in section 4, there exist on the manifold M vector fields X such that $X|_V = Y|_V$. Let X_1 and X_2 denote two such fields on M, and let h denote a function that is smooth on M, that is equal to unity at the point p, and that vanishes outside the neighborhood V. Then, $h \cdot (X_1 - X_2) = 0$ on M and, hence,

$$h \cdot (\omega(X_1) - \omega(X_2)) = 0 \quad \text{on } M.$$

This shows that the functions $\omega(X_1)$ and $\omega(X_2)$ coincide at the point p. Since the point p was chosen arbitrarily in W, this proves that the formula

$$\omega|_W(Y)(p) = \omega(X)(p),$$

where X is an arbitrary field on M such that $X|_V = Y|_V$, defines uniquely a function $\omega|_W(Y)$ [obviously, a smooth function] on W. One can show without difficulty that the correspondence $Y \to \omega|_W(Y)$ is $6(W)$-linear, that is, that it is a linear differential form ω_W on W. We shall say that this form is a *restriction* of the form ω to the open submanifold W and, when there is no danger of confusion, we shall use the symbol ω for it also.

Just as for vector fields, the mapping of the restriction $\omega \to \omega|_W$ is locally epimorphic; that is,

For any open set V having a compact closure contained in W and for any linear differential form θ on W, there exists a linear differential form ω on M such that

$$\theta|_V = \omega|_V.$$

This form ω is determined by the formula

$$\omega(X) = \begin{cases} h\theta(X|_W) & \text{on} \quad W, \\ 0 \text{ outside} & W, \end{cases}$$

where h is a function that is smooth on M, that is equal to one on $\overline{V}$, and that vanishes outside some open set whose closure is contained in W.

In the particular case in which W is a coordinate neighborhood U, it is easy to describe all the linear differential forms on W. Let $X_1, \ldots, X_m$ be an arbitrary basis of the module $6^1(U)$. Obviously, for arbitrary $i = 1, \ldots, m$, the relations

$$\omega^i(X_j) = \delta^i_j, \qquad j = 1, \ldots, m$$

define uniquely on U some linear differential form ω^i. The forms $\omega^1, \ldots, \omega^m$ are linearly independent and they have the property that, for an arbitrary vector field $X \in 6^1(U)$,

$$X^i = \omega^i(X), \qquad i = 1, \ldots, m,$$

where the X^i are the components of the field X relative to the basis $X_1, \ldots, X_m$. From this it follows directly that

The forms $\omega^1, \ldots, \omega^m$ constitute a basis for the module $6_1(U)$, and

$$\omega = \omega(X_i)\omega^i$$

for an arbitrary form $\omega \in 6_1(U)$.

We shall call this basis the *dual* of the basis $X_1, \ldots, X_m$.

We shall call the functions $\omega_i = \omega(X_i)$ the *components* of the form ω relative to the basis $X_1, \ldots, X_m$ (or to the basis $\omega^1, \ldots, \omega^n$).

If the basis $X_1, \ldots, X_m$ is holonomic, that is, if

$$X_1 = \frac{\partial}{\partial x^1}, \ldots, X_m = \frac{\partial}{\partial x^m},$$

we shall also say that the basis $\omega^1, \ldots, \omega^m$ is holonomic. In this case, we shall usually denote the forms $\omega^1, \ldots, \omega^m$ by the symbols

$$dx^1, \ldots, dx^m.$$

We shall sometimes call the components $\omega\left(\frac{\partial}{\partial x_i}\right)$ of the form ω relative to the basis $dx^1, \ldots, dx^m$ its ***coefficients*** (in the system of local coordinates $x^1, \ldots, x^m$).

An important example of a linear differential form on the manifold M is obtained by setting

$$df(X) = X(f).$$

where f is an arbitrary smooth function on M and X is an arbitrary field in $\mathfrak{G}^1(M)$. Obviously, this formula does not define on M a linear differential form df. We shall call the form df the ***differential*** of the function f.

By definition, for an arbitrary function f that is smooth on M (or even on the coordinate neighborhood U)

$$df\left(\frac{\partial}{\partial x_i}\right) = \frac{\partial f}{\partial x^i},$$

so that, just as in elementary calculus,

$$df = \frac{\partial f}{\partial x^i} dx^i \quad \text{on} \quad U.$$

In particular, we see that the forms $dx^1, \ldots, dx^m$ of a holonomic basis are simply the differentials of the local coordinates $x^1, \ldots, x^m$.

8. Covectors

We shall refer to linear functionals defined on the space M_p, that is, the linear mappings of this space into the field R, as ***covectors*** of the manifold M at the point p. They constitute an m-dimensional linear space M_p^* dual to the space M_p. We shall also denote the space M_p^* by the symbol $\mathfrak{G}_1(p)$.

It is easy to see that, for an arbitrary form $\omega \in \mathfrak{G}_1(M)$ and an arbitrary field $X \in \mathfrak{G}^1(M)$, the equation $X_p = 0$ implies the equation $\omega(X)(p) = 0$. To see this, note that, in the coordinate neighborhood U of the point p,

$$\omega(X) = \omega\left(\frac{\partial}{\partial x^i}\right) X x^i$$

and, hence,

$$\omega(X)(p) = \omega\left(\frac{\partial}{\partial x^i}\right)(p)\, X_p x^i(p) = 0.$$

From this it follows that the formula

$$\omega_p(A) = \omega(X)(p),$$

where A is an arbitrary vector in the space M_q and X is a vector field on M such that $X_p = A$, unambigously defines on M_p a covector ω_p. Here,

For an arbitrary covector $\alpha \in 6_1(p)$, *there exists a form* ω *on M such that* $\omega_p = \alpha$.

Proof: Because of the local epimorphism of the restriction mapping, it will be sufficient to construct the form ω on some coordinate neighborhood U of the point p. On the other hand, if we set

$$\omega = \alpha_i\, dx^i, \text{where } \alpha_i = \alpha\left(\left(\frac{\partial}{\partial x^i}\right)_p\right),$$

we shall obviously obtain on U a form ω such that $\omega_p = \alpha$.

The function $p \to \omega_p$ that assigns to each point $p \in M$ the covector ω_p obviously satisfies the following "smoothness condition":

For an arbitrary vector field $X \in 6^1(M)$, *the real function* $f(p) = \omega_p(X_p)$ *is smooth on M.*

It is easy to see that, conversely,

Any function $p \to \omega_p$ *that assigns to each point* $p \in M$ *a convector* $\omega_p \in 6_1(p)$ *and satisfies this smoothness condition satisfies some linear differential form* ω*.*

This form ω is determined by the formula

$$\omega(X)(p) = \omega_p(X_p), \quad X \in 6^1(M), \quad p \in M.$$

In view of this, the linear differential forms are also called covector fields.

As has been stated, the space $\overset{*}{M}_p$ has dimension m. A basis for it is, for example, the set of convectors

$$(dx^1)_p, \ \ldots, \ (dx^m)_p,$$

where $x^1, \ldots, x^m$ are an arbitrary system of local coordinates at the point p. An arbitrary covector $\alpha \in \overset{*}{M}_p$ can be decomposed with respect to this basis according to the formula

$$\alpha = \alpha_i (dx^i)_p, \text{where } \alpha_i = \alpha\left(\left(\frac{\partial}{\partial x^i}\right)_p\right).$$

Remark: An arbitrary smooth function f on M defines at every point $p \in M$ a covector $(df)_p$. On the other hand, if we consider the function f as a smooth mapping $M \to R$, we can determine the differential $df_p : M_p \to R_{f(p)}$, where $R_{f(p)}$ is the tangent space of the space R at the point $f(p)$. It is easy to see that, if we identify each vector $a\left(\frac{\partial}{\partial t}\right)_{f(p)} \in R_{f(p)}$ with the number $a \in R$, the differential df_p

will coincide with the covector $(df)_p$. In this sense, the definition of the differential of a function is consistent with the definition of the differential of a smooth mapping.

In conclusion, let us consider an arbitrary smooth mapping Φ of some manifold N into the manifold M. Let ω denote a linear differential form on M. At every point $p \in N$, we can define a covector $(\Phi^*\omega)_p$ by setting

$$(\Phi^*\omega)_p(A) = \omega_p(d\Phi_p(A))$$

for an arbitrary vector $A \in N_p$. It is easy to see that the mapping $p \to (\Phi^*\omega)_p$ satisfies the smoothness condition and hence defines on N a linear differential form $\Phi^*(\omega)$. It is also obvious that the mapping so defined

$$\Phi^* : \mathfrak{G}_1(M) \to \mathfrak{G}_1(N)$$

is a homomorphism of $\mathfrak{G}(M)$-modules (if we consider the $\mathfrak{G}(N)$-module $\mathfrak{G}_1(N)$ as an $\mathfrak{G}(M)$-module on the basis of the mapping $\mathfrak{G}(M) \to \mathfrak{G}(N)$ that assigns to each function $f \in \mathfrak{G}(M)$ the function $f \circ \Phi \in \mathfrak{G}(N)$).

9. Tensor fields

Suppose that r and s are arbitrary nonnegative integers. Let us consider functions

$$T(\omega^1, \ldots, \omega^r; X_1, \ldots, X_s)$$

of $r+s$ variables, the first r of which are linear differential forms (covector fields) $\omega^1, \ldots, \omega^r$ on M and the last s of which are vector fields $X_1, \ldots, X_s$ on M. Suppose that the range of such a function T is contained in the algebra $\mathfrak{G}(M)$. If such a function is $\mathfrak{G}(M)$-linear in each argument, we shall call it a *tensor field* of the type (r, s) on the manifold M.

We shall denote the set of all tensor fields of the type (r, s) on the manifold M by the symbol $\mathfrak{G}_s^r(M)$. Obviously, the construction of an $\mathfrak{G}(M)$-module is introduced in a natural manner into this set.

By definition, tensor fields of the type $(0, 0)$ are simply functions that are smooth on M, so that

$$\mathfrak{G}_0^0(M) = \mathfrak{G}(M),$$

and tensor fields of the type $(0, 1)$ are simply linear differential forms, so that

$$\mathfrak{G}_1^0(M) = \mathfrak{G}_1(M).$$

Furthermore, an arbitrary vector field $X \in \mathfrak{G}^1(M)$ can be regarded as a tensor field of the type $(1,0)$ if we define its value $X(\omega)$ on an arbitrary form $\omega \in \mathfrak{G}_1(M)$ by

$$X(\omega) = \omega(X).$$

Since $\omega(X) = 0$ for all forms $\omega \in 6_1(M)$ only when $X = 0$, it follows that the module $6^1(M)$ is embedded in the module $6_0^1(M)$. It turns out, in fact, that these modules coincide:

$$6_0^1(M) = 6^1(M), \tag{1}$$

that is,

The set of all vector fields and the set of all tensor fields of the type (1, 0) *are the same.*

We shall prove this important assertion somewhat later, after we have studied in greater detail the structure of arbitrary tensor fields.

Suppose that T is an arbitrary tensor field of the type (r, s) on M and suppose that W is an open submanifold of the manifold M. Let us show that

If at least one of the arguments $\omega^1, \ldots, \omega^r, X_1, \ldots, X_s$ *of the field* T *vanishes on* W,[1] *then*

$$T(\omega^1, \ldots, \omega^r, X_1, \ldots, X_s) = 0$$

on W.

Proof: Let p denote an arbitrary point in W and let V denote a neighborhood of p with compact closure $\bar{V} \subset W$. As we know, there exists on M a smooth function f that is equal to unity at p and that vanishes outside V. Suppose, for definiteness, that $\omega^1 = 0$ on W. Then, $f\omega^1 = 0$ on M and, hence,

$$T(f\omega^1, \omega^2, \ldots, \omega^r; X_1, \ldots, X_s) = 0.$$

But

$$T(f\omega^1, \omega^2, \ldots, \omega^r; X_1, \ldots, X_s) = fT(\omega^1, \ldots, \omega^r; X_1, \ldots, X_s)$$

and, therefore,

$$T(\omega^1, \ldots, \omega^r; X_1, \ldots, X_s)(p) =$$
$$= T(f\omega^1, \omega^2, \ldots, \omega^r; X_1, \ldots, X_s)(p) = 0,$$

since $f(p) = 1$. Consequently, $T(\omega^1, \ldots, \omega^r; X_1, \ldots, X_s) = 0$ on W.

Suppose now that $\theta^1, \ldots, \theta^r$ are arbitrary forms and $Y_1, \ldots, Y_s$ are arbitrary vector fields on an open submanifold W. For an arbitrary point $p \in W$, let us consider on M the forms $\omega^1, \ldots, \omega^r$ and the fields $X_1, \ldots, X_s$ that coincide in some neighborhood of the point p with forms $\theta^1, \ldots, \theta^r$ and fields $Y_1, \ldots, Y_s$, respectively. From the assertion that we have just proved, it follows immediately (cf. the corresponding reasoning above regarding vector fields and differential forms) that the formula

---

[1] That is, if its restriction to W vanishes identically.

$$T|_W(\theta^1, \ldots, \theta^r; Y_1, \ldots, Y_s)(p) = T(\omega^1, \ldots, \omega^r; X_1, \ldots, X_s)(p)$$

defines unambiguously on W a smooth function

$$T|_W(\theta^1, \ldots, \theta^r; Y_1, \ldots, Y_s).$$

Obviously, the mapping

$$(\theta^1, \ldots, \theta^r, Y_1, \ldots, Y_s) \to T|_W(\theta^1, \ldots, \theta^r; Y_1, \ldots, Y_s)$$

is $6(M)$-linear in $\theta^1, \ldots, \theta^r, Y_1, \ldots, Y_s$; that is, it is a tensor field of the type (r, s) on W. We shall call the field $T|_W$ the restriction of the field T to the open submanifold W and, when there is no danger of misunderstanding, we shall denote it too by the symbol T.

Obviously, the mapping $T \to T|_W$ is an $6(M)$-homomorphism of the $6(M)$-module $6_s^r(M)$ into the $6(M)$-module $6_s^r(W)$ [which, by virtue of the restriction mapping $6(M) \to 6(W)$, is considered as an $6(M)$-module]. Here, in analogy with the case of vector fields and linear differential forms, this homomorphism is locally epimorphic; that is,

For any open set V with compact closure $V \subset W$ and for any field $S \in 6_s^r(W)$, there exists a field $T \in 6_s^r(M)$ such that

$$T|_V = S|_V.$$

In particular, let us consider the restriction $T|_U$ of the field $T \in 6_s^r(M)$ on some coordinate neighborhood U. Let $Y_1, \ldots, Y_m$ denote an arbitrary basis of the module $6^1(U)$ and let $\theta_1, \ldots, \theta_m$ denote the corresponding basis of the module $6_1(U)$ [cf. section 7]. Then, for arbitrary forms $\omega^1, \ldots, \omega^r \in 6_1(U)$ and arbitrary fields $X_1, \ldots, X_s \in 6^1(U)$

$$
\begin{aligned}
T(\omega^1, \ldots, \omega^r; X_1, \ldots, X_s) &= \\
&= T_{l_1 \cdots l_s}^{k_1 \cdots k_r} \omega_{k_1}^1 \ldots \omega_{k_r}^r X_1^{l_1} \ldots X_s^{l_s} \quad \text{on} \quad U,
\end{aligned} \tag{2}
$$

where

$$T_{l_1 \cdots l_s}^{k_1 \cdots k_r} = T\left(\theta^{k_1}, \ldots, \theta^{k_r}; Y_{l_1}, \ldots, Y_{l_s}\right), \tag{3}$$

and the ω_k^i (for $i = 1, \ldots, r$ and $k = 1, \ldots, m$) and the X_j^l (for $j = 1, \ldots, s$ and $l = 1, \ldots, m$) are the components of the forms ω^i and fields X_j relative to the basis $Y_1, \ldots, Y_m$. We shall call the function (3) the components of the tensor field T or U (relative to the basis $Y_1, \ldots, Y_m$). They completely determine the field T on the neighborhood U.

In particular, if a field T is of the type $(1, 0)$, its components are the functions $T^k = T(\theta^k)$. Let us define on U a vector field X_U by

$$X_U = T^k Y_k.$$

Then, for an arbitrary form $\omega \in \mathfrak{G}_1(U)$,

$$T(\omega) = \omega(Y_k) T(\theta^k) = \omega(T^k Y_k) = \omega(X_U) \quad \text{on} \quad U,$$

This shows that the tensor field T coincides on U with the vector field X_U. From this it follows, in particular, that the field X_U is unambiguously defined (does not depend on the choice of the basis $Y_1, \ldots, Y_m$). Therefore, by setting

$$X_p = (X_U)_p,$$

where p is an arbitrary point in M and U is an arbitrary coordinate neighborhood of the point p, we obtain on M an unambiguously defined function $p \to X_p$ that obviously satisfies the smoothness condition, i.e., is a vector field. Since $T(\omega) = \omega(X_U)$ on U, we have

$$T(\omega)(p) = \omega(X_U)(p) = \omega_p((X_U)_p) = \omega_p(X_p) = \omega(X)(p),$$

that is,

$$T(\omega) = \omega(X) \quad \text{on } M.$$

In other words, $T = X$ on M. This completes the proof of formula (1).

10. Tensors. The multiplications of tensors and tensor fields

Let p denote an arbitrary point of a smooth manifold M. Consider a functional of the form

$$L(\alpha^1, \ldots, \alpha^r; A_1, \ldots, A_s),$$

that is linear in each of its arguments, where the first r arguments range over the space $\mathfrak{G}_1(p)$ and the last s over the space $\mathfrak{G}^1(p)$. We shall call such a functional a tensor of type (r, s) at the point p. All tensors of type (r, s) at the point p obviously generate a linear space $\mathfrak{G}^r_s(p)$. Clearly,

$$\mathfrak{G}^1_0(p) = \mathfrak{G}^1(p), \ \mathfrak{G}^0_1(p) = \mathfrak{G}_1(p).$$

On the other hand, $\mathfrak{G}^0_0(p) = R$ [and not $\mathfrak{G}(p)$].

For an arbitrary tensor field $T \in \mathfrak{G}^r_s(M)$, it follows easily from formula (2) of section 9 that vanishing of any one of its arguments $\omega^1, \ldots, \omega^r, X_1, \ldots, X_s$ at the point p implies

$$T(\omega^1, \ldots, \omega^r; X_1, \ldots, X_s)(p) = 0.$$

Therefore, the formula

$$T_p(\alpha^1, \ldots, \alpha^r; A_1, \ldots, A_s) = T(\omega^1, \ldots, \omega^r; X_1, \ldots, X_s)(p),$$

where the forms $\omega^1, \ldots, \omega^r$ and the fields $X_1, \ldots, X_s$ are chosen in such a way that

$$(\omega^i)_p = \alpha^i, \ (X_j)_p = A_j,$$

defines a tensor T_p of type (r, s) unambiguously at the point p.

Thus, every tensor field $T \in \mathfrak{T}_s^r(M)$ defines a function $p \to T_p$ that maps each point $p \in M$ into some tensor $T_p \in \mathfrak{T}_s^r(p)$. Obviously, this function satisfies the following smoothness condition:

For arbitrary forms $\omega^1, \ldots, \omega^r \in \mathfrak{T}_1(M)$ *and arbitrary vector fields* $X_1, \ldots, X_s \in \mathfrak{T}^1(M)$, *the real function*

$$f(p) = T_p((\omega^1)_p, \ldots, (\omega^r)_p; (X_1)_p, \ldots, (X_s)_p), \qquad p \in M,$$

is smooth on M.

It is easy to see that, conversely,

An arbitrary function $p \to T_p$ *that maps each point* $p \in M$ *into some tensor* $T_p \in \mathfrak{T}_s^r(p)$ *and that satisfies this smoothness condition corresponds to some tensor field* $T \in \mathfrak{T}_s^r(M)$.

This field T is given by the formula

$$T(\omega^1, \ldots, \omega^r; X_1, \ldots, X_s)(p) = \\ = T_p((\omega^1)_p, \ldots, (\omega^r)_p; (X_1)_p, \ldots, (X_s)_p).$$

If we choose some basis $B_1, \ldots, B_m$ in the space $M_p = \mathfrak{T}^1(p)$ and if we construct for the space $M_p^* = \mathfrak{T}_1(p)$ the dual basis $\beta^1, \ldots, \beta^m$ (that is, the basis satisfying the relations $\beta^j(B_i) = \delta_i^j$), we see that the value $L(\alpha^1, \ldots, \alpha^r; A_1, \ldots, A_s)$ of an arbitrary tensor $L \in \mathfrak{T}_s^r(p)$ on arbitrary covectors $\alpha^1, \ldots, \alpha^r \in \mathfrak{T}_1(p)$ and vectors $A_1, \ldots, A_s \in \mathfrak{T}^1(p)$ is given by the formula

$$L(\alpha^1, \ldots, \alpha^r; A_1, \ldots, A_s) = L_{l_1 \ldots l_s}^{k_1 \ldots k_r} \alpha_{k_1}^1 \ldots \alpha_{k_r}^r A_1^{l_1} \ldots A_s^{l_s}, \tag{1}$$

where $\alpha_k^i = \alpha^i(B_k)$ and $A_j^l = \beta^l(A_j)$ are the components of the covectors α^i and vectors A_j and where

$$L_{l_1 \ldots l_s}^{k_1 \ldots k_r} = L\left(\beta^{k_1}, \ldots, \beta^{k_r}; B_{l_1}, \ldots, B_{l_s}\right)$$

are the *components* of the tensor L relative to the basis $B_1, \ldots, B_m$. Clearly, for an arbitrary choice of the m^{r+s} numbers $L_{l_1 \ldots l_s}^{k_1 \ldots k_r}$, formula (1) defines a tensor $L \in \mathfrak{T}_s^r(p)$ with components $L_{l_1 \ldots l_s}^{k_1 \ldots k_r}$.

Now, let L and K denote two tensors of types (r, s) and (u, v), respectively, at a point p. We define the tensor $L \otimes K$, of the type $(r+u, s+v)$ at the point p by

$$(L \otimes K)(\alpha^1, \ldots, \alpha^{r+u}; A_1, \ldots, A_{s+v}) = \\ = L(\alpha^1, \ldots, \alpha^r; A_1, \ldots, A_s) K(\alpha^{r+1}, \ldots, \alpha^{r+u}; A_{s+1}, \ldots, A_{s+v})$$

for arbitrary covectors $\alpha^1, \ldots, \alpha^{r+u} \in \mathfrak{T}_1(p)$ and arbitrary vectors $A_1, \ldots, A_{s+v} \in \mathfrak{T}^1(p)$. Obviously, the components of the tensor $L \otimes K$ are given by the formula

$$(L \otimes K)_{l_1 \ldots l_{s+v}}^{k_1 \ldots k_{r+u}} = L_{l_1 \ldots l_s}^{k_1 \ldots k_r} K_{l_{s+1} \ldots l_{s+v}}^{k_{r+1} \ldots k_{r+u}}.$$

We shall say that the tensor $L \otimes K$ is the *product* of the tensors L and K. It is easy to verify that

Relative to the product $\otimes$, the direct sum

$$\mathrm{O}(p) = \sum_{r,\, s=0}^{\infty} \mathfrak{G}_s^r(p)$$

of linear spaces $\mathfrak{G}_s^r(p)$ is an associative algebra over the field R.

Suppose now that T and S are arbitrary tensors fields on a manifold M. We shall define their product $T \otimes S$ by setting, for an arbitrary point $p \in M$,

$$(T \otimes S)_p = T_p \otimes S_p.$$

It is easy to see that this formula does indeed define a tensor field, that is, that the mapping $p \to (T \otimes S)_p$ satisfies the smoothness condition mentioned above. It is also clear that

Relative to the product $\otimes$, the direct sum

$$\mathrm{O}(M) = \sum_{r,\, s=0}^{\infty} \mathfrak{G}_s^r(M)$$

of the modules $\mathfrak{G}_s^r(M)$ is an associative algebra over the ring $\mathfrak{G}(M)$.

We note that, for tensors of type $(0, 0)$, that is, for functions that are smooth on M, the multiplication $\otimes$ coincides with the usual multiplication of functions.

Let us look, in particular, at the algebra $\mathrm{O}(U)$, where U is a coordinate neighborhood of the manifold M. Suppose that $x^1, \ldots, x^m$ are local coordinates defined in the neighborhood U. Obviously, for an arbitrary tensor field $T \in \mathfrak{G}_s^r(U)$, the field

$$T_{l_1 \ldots l_s}^{k_1 \ldots k_r} dx^{l_1} \otimes \ldots \otimes dx^{l_s} \otimes \frac{\partial}{\partial x^{k_1}} \otimes \cdots \otimes \frac{\partial}{\partial x^{k_r}},$$

where the $T_{l_1 \ldots l_s}^{k_1 \ldots k_r}$ are the components of the field T in the local coordinate system $x^1, \ldots, x^m$ (that is, relative to the basis $\frac{\partial}{\partial x^1}, \ldots, \frac{\partial}{\partial x^m}$), has the same components $T_{l_1 \ldots l_s}^{k_1 \ldots k_r}$ as the field T. This means that

$$T = T_{l_1 \ldots l_s}^{k_1 \ldots k_r} dx^{l_1} \otimes \ldots \otimes dx^{l_s} \otimes \frac{\partial}{\partial x^{k_1}} \otimes \ldots \otimes \frac{\partial}{\partial x^{k_r}}. \tag{2}$$

Thus,

The algebra $\mathrm{O}(U)$ is generated over the ring $\mathfrak{G}(U)$ by the forms $dx^1, \ldots, dx^m$ and the fields $\frac{\partial}{\partial x^1}, \ldots, \frac{\partial}{\partial x^m}$; each tensor $T \in \mathfrak{G}_s^r(U)$ can be expressed in a unique manner in terms of these generators in accordance with formula (2).

Obviously, an analogous assertion holds for the algebras $\mathfrak{O}(p)$.

In conclusion, let us consider an arbitrary smooth mapping Φ of a manifold N into a manifold M. Let T denote a tensor field of type $(0, s)$ on the manifold M. At every point $p \in N$, we define a tensor $(\Phi^*T)_p$ of type $(0, s)$ by

$$(\Phi^*T)_p\,(A_1, \ldots, A_s) = T_{\Phi(p)}(d\Phi_p(A_1), \ldots, d\Phi_p(A_s)),$$

where $A_1, \ldots, A_s$ are arbitrary vectors in N_p. One can easily verify that the mapping $p \to (\Phi^*T)_p$ satisfies the smoothness condition and hence determines a tensor field Φ^*T on the manifold N. Obviously, the mapping

$$\Phi^* : \mathfrak{T}_s^0(M) \to \mathfrak{T}_s^0(N)$$

is a homomorphism of $\mathfrak{T}(M)$-modules (if, on the basis of the mapping $\mathfrak{T}(M) \to \mathfrak{T}(N)$ which assigns to each function $f \in \mathfrak{T}(M)$ the function $f \circ \Phi \in \mathfrak{T}(N)$, we regard $\mathfrak{T}(N)$-module $\mathfrak{T}_s^0(N)$ as an $\mathfrak{T}(M)$-modulus).

Obviously, for tensor fields of type $(0, 1)$, that is, for linear differential forms, the mapping Φ^* defined above coincides with the mapping Φ^* defined in section 8. The mapping Φ^* cannot be defined for fields of type (r, s) with $r \neq 0$.

11. The contraction of tensors and tensor fields

Suppose that i and j are integers satisfying the inequalities $1 \leqslant i \leqslant s$ and $1 \leqslant j \leqslant r$. In some tensor

$$L\,(\alpha^1, \ldots, \alpha^r;\ A_1, \ldots, A_s) \in \mathfrak{T}_s^r\,(p)$$

let us fix all the arguments $\alpha^1, \ldots, \alpha^r, A_1, \ldots, A_s$ except the ith vector argument A_i. This gives us a linear functional defined on the space $\mathfrak{T}^1(p)$, that is, a linear covector in $\mathfrak{T}_1(p)$. Of course, this covector depends on the covectors $\alpha^1, \ldots, \alpha^r$ and the vectors $A_1, \ldots, \widehat{A}_i, \ldots, A_s$, where the circumflex accent indicates that the corresponding vector should be omitted. In particular, it depends on the covector α^j and the dependence is obviously linear. Consequently, if we fix all the covectors $\alpha^1, \ldots, \alpha^r$ except the jth and all the vectors $A_1, \ldots, A_s$ except the ith, we obtain from the tensor L a linear transformation of the space $M_p^* = \mathfrak{T}_1(p)$. Let

$$(C_j^i L)(\alpha^1, \ldots, \widehat{\alpha}^j, \ldots, \alpha^r;\ A_1, \ldots, \widehat{A}_i, \ldots, A_s)$$

be the trace of this linear transformation. This trace is a linear function of the $r-1$ covectors $\alpha^1, \ldots, \widehat{\alpha}^j, \ldots, \alpha^r$ and the $s-1$ vectors $A_1, \ldots, \widehat{A}_i, \ldots, A_s$; that is, it is a tensor of the type $(r-1, s-1)$ at the point p. We shall say that this tensor $C_j^i L$ is obtained from the tensor L by contraction with respect to the ith upper and jth lower indices. Its components $(C_j^i L)_{i_1 \ldots i_{s-1}}^{k_1 \ldots k_{r-1}}$ are expressed in terms of the components $L_{l_1 \ldots l_s}^{k_1 \ldots k_r}$ of the tensor L in accordance with the formula

$$(C_j^i L)_{l_1 \ldots l_{s-1}}^{k_1 \ldots k_{r-1}} = L_{l_1 \ldots l_{j-1} h l_j \ldots l_{s-1}}^{k_1 \ldots k_{i-1} h k_i \ldots k_{r-1}}.$$

We shall call the mapping constructed in this way

$$C_j^i : 6_s^r(p) \to 6_{s-1}^{r-1}(p)$$

a *contraction mapping.*

It is easy to see that, for an arbitrary tensor field $T \in 6_s^r(M)$, the formula

$$(C_j^i T)_p = C_j^i (T_p), \qquad p \in M,$$

defines on M a tensor field $C_j^i T$ of type $(r-1, s-1)$, and the mapping obtained

$$C_j^i : 6_s^r(M) \to 6_{s-1}^{r-1}(M)$$

is a homomorphism of $6(M)$-modules. We shall also call it a *contraction mapping.*

The case $r = s$ is of particular interest. In this case, if we apply the contraction mapping r times to the field T, we obtain a field of type $(0, 0)$, that is, a function that is smooth on M. We shall say that this function was obtained from the field T as a result of a complete contraction. Clearly, the result of a *complete contraction* depends, in general, on the pair of indices with respect to which each of the r partial contractions was performed.

In particular, for an arbitrary tensor field T of type (r, s), let us consider the field

$$S = T \otimes \omega^1 \otimes \ldots \otimes \omega^r \otimes X_1 \otimes \ldots \otimes X_s$$

(of type $(r+s, r+s)$, where $\omega^1, \ldots, \omega^r \in 6_1(M)$ and $X_1 \ldots, X_s \in 6^1(M)$. Suppose that $p \in M$. If we denote by ω_j^i and X_l^k the components of the covectors $(\omega^l)_p$ and the vectors $(X_l)_p$ relative to some basis for the space M_p, we see that the components of the tensor S_p are expressed by the formula

$$(S_p)_{l_1 \ldots l_{r+s}}^{k_1 \ldots k_{r+s}} = (T_p)_{l_1 \ldots l_s}^{k_1 \ldots k_r} \omega_{l_{s+1}}^1 \ldots \omega_{l_{r+s}}^r X_1^{k_{r+1}} \ldots X_s^{k_{r+s}},$$

where the $(T_p)_{l_1 \ldots l_r}^{k_1 \ldots k_r}$ are the components of the tensor T_p. From this it follows that, by contracting the tensor S_p with respect to the indices $(1, s+1), \ldots, (r, r+s), (r+1, 1), \ldots, (r+s, s)$, we obtain the number

$$(T_p)_{l_1 \ldots l_s}^{k_1 \ldots k_r} \omega_{k_1}^1 \ldots \omega_{k_r}^r X_1^{l_1} \ldots X_s^{l_s} =$$
$$= T_p\big((\omega^1)_p, \ldots, (\omega^r)_p; (X_1)_p \ldots (X_s)_p\big) =$$
$$= T(\omega^1, \ldots, \omega^r; X_1, \ldots, X_s)(p).$$

Thus,

As a result of the complete contraction described above of the field

$$T \otimes \omega^1 \otimes \ldots \otimes \omega^r \otimes X_1 \otimes \ldots \otimes X_s$$

we obtain the function $T(\omega^1, \ldots, \omega^r; X_1, \ldots, X_s)$.

12. Curves and surfaces

Just as above, let M denote an arbitrary smooth manifold. We shall refer to an arbitrary continuous mapping $\gamma\colon I \to M$ of an open interval $I \subset R$ into the manifold M as a *continuous curve* on M. We shall say that the curve γ is *smooth* at the point $t \in I$ if the mapping γ is smooth at t and we shall say that it is *regular* at t if the mapping γ is smooth at t and its differential $d\gamma_t$ is nonzero at that point. We shall say that a continuous curve γ is *piecewise smooth* (resp. *piecewise regular*) if it is smooth (resp. regular) at all but a finite number of points in I. If a curve is smooth (resp. regular) at all points of the interval I, we shall call it a smooth (resp. regular) curve. If a smooth curve $\gamma\colon I \to M$ is a constant mapping, that is, if $\gamma(t) = p$, where p is a fixed point in the manifold M, for all $t \in I$, we shall call it a *degenerate curve*.

We shall refer to the restriction $\gamma|_C$ of a curve γ to a closed nondegenerate interval $C \subset I$ (that is, not consisting of a single point) as a *segment* of the curve γ. We shall call the segment $\gamma|_C$ finite if the interval C is finite.

For an arbitrary smooth function f on M, the curve γ defines a real function

$$f(t) = f(\gamma(t)), \quad t \in I.$$

This function is smooth if the curve γ is smooth.

In particular, suppose that U is a coordinate neighborhood for which there exists an interval $J \subset I$ such that $\gamma(t) \in U$ for points $t \in J$. Then, the functions

$$x^i(t) = x^i(\gamma(t)), \qquad i = 1, \ldots, m,$$

where $x^1, \ldots, x^m$ are local coordinates in the neighborhood U, are defined on J. These functions determine the curve $\gamma|_J$ unambiguously. We shall say that they define the curve γ (or, more precisely, the curve $\gamma|_J$) parametrically in the neighborhood U. As one can easily see,

A curve $\gamma(t)$ is a smooth curve if and only if, for an arbitrary system of local coordinates $x^1, \ldots, x^m$, the functions $x^1(t), \ldots, x^m(t)$ defining the curve $\gamma(t)$ parametrically are smooth (that is, infinitely differentiable) functions of the parameter t.

Suppose now that to each number $t \in I$ is assigned a tensor $T(t) \in \mathfrak{T}^r_s(\gamma(t))$ of type $(r, s) \neq (0, 0)$. We shall say that the function $T(t)$

is a tensor field on the curve γ if, for arbitrary differential forms $\omega^1, \ldots, \omega^r \in \mathfrak{G}_1(M)$ and arbitrary vector fields $X_1, \ldots, X_s \in \mathfrak{G}^1(M)$, the real function

$$f(t) = T(t)\left((\omega^1)_{\gamma(t)} \ldots (\omega^r)_{\gamma(t)}; (X_1)_{\gamma(t)}, \ldots, (X_s)_{\gamma(t)}\right)$$

is smooth on I. If $J \subset I$ is an interval such that $\gamma(J) \subset U$, where U is a coordinate neighborhood of the manifold M, then, for arbitrary $t \in J$,

$$T(t) = T_{l_1 \ldots l_s}^{k_1 \ldots k_r}(t)\left(dx^{l_1}\right)_{\gamma(t)} \otimes \ldots \otimes \left(dx^{l_s}\right)_{\gamma(t)} \otimes \otimes \left(\frac{\partial}{\partial x^{k_1}}\right)_{\gamma(t)} \otimes \ldots \otimes \left(\frac{\partial}{\partial x^{k_r}}\right)_{\gamma(t)}, \tag{1}$$

where the $T_{l_1 \ldots l_s}^{k_1 \ldots k_r}(t)$ are smooth functions on J [the *components* of the field $T(t)$].

We shall be particularly interested in the vector fields $X(t)$, that is, tensor fields of type $(1, 0)$. For such a field,

$$X(t) = X^i(t)\left(\frac{\partial}{\partial x^i}\right)_{\gamma(t)}, \quad t \in J, \tag{2}$$

where the $X^i(t) = (Xx^i)(\gamma(t))$ are smooth functions on J.

An example of a vector field defined on a smooth curve γ is the field

$$\dot{\gamma}(t) = d\gamma_t\left(\frac{\partial}{\partial t}\right).$$

We shall call this field the vector field of tangents to the curve γ. Obviously,

A curve γ *is regular at a point* $t \in I$ *if and only if the vector* $\dot{\gamma}(t)$ *is nonzero.*

Since the operation $\partial/\partial t$ coincides with ordinary differentiation with respect to t in the case of real functions of t, it follows that, for the derivatives $\dot{x}^i(t) = \frac{dx^i(t)}{dt}$ of the functions $x^i(t)$ defining a smooth curve γ parametrically in some local coordinate system,

$$\dot{x}^i(t) = \frac{\partial}{\partial t}(x^i(\gamma(t))) = d\gamma_t\left(\frac{\partial}{\partial t}\right)x^i = \dot{\gamma}(t)\,x^i,$$

from which it follows that

$$\dot{\gamma}(t) = \dot{x}^i(t)\left(\frac{\partial}{\partial x^i}\right)_{\gamma(t)}. \tag{3}$$

Thus,

The components of the field $\dot{\gamma}(t)$ *are the functions* $\dot{x}^i(t)$.

One consequence of this, among others, is the fact that, for an arbitrary function f that is smooth on M, the derivative $\dot{f}(t)=\frac{df(t)}{dt}$ of the function $f(t)=f(\gamma(t))$ can be calculated from the usual chain rule for differentiating a composite function:

$$\frac{df(t)}{dt}=\left(\frac{\partial f}{\partial x^i}\right)_{\gamma(t)}\frac{dx^i(t)}{dt},$$

where $x^1, \ldots, x^m$ are local coordinates. This is true because

$$\frac{df(t)}{dt}=\frac{\partial}{\partial t}f(\gamma(t))=\dot{\gamma}(t)f=$$
$$=\dot{x}^i(t)\left(\frac{\partial}{\partial x^i}\right)_{\gamma(t)}f=\left(\frac{\partial f}{\partial x^i}\right)_{\gamma(t)}\frac{dx^i(t)}{dt}.$$

Suppose now that X is an arbitrary vector field on a manifold M. We shall call curve $\gamma: I\to M$ an ***integral curve*** of the field $X\in\mathfrak{G}^1(M)$ if this curve is regular and $\dot{\gamma}(t)=X_{\gamma(t)}$ for every $t\in I$.

In view of equation (3), this condition in an arbitrary system of local coordinates $x^1, \ldots, x^m$ can be expressed in the form of a system of differential equations

$$\dot{x}^i(t)=X^i(x^i(t), \ldots, x^m(t)), \tag{4}$$

where the X^i are real functions, of m variables, defined by

$$Xx^i=X^i(x^1, \ldots, x^m) \text{ on } U.$$

If an integral curve $\gamma(t)$ passes through a point $p\in M$ (that is, if $\gamma(t_0)=p$ for some $t_0\in I$), then $X_p\neq 0$ (since $\dot{\gamma}(t_0)\neq 0$). Conversely, if $X_p\neq 0$, then the right-hand members of the system (4) are nonzero and hence this system has (let us say, in a neighborhood of the point $t=0$) a solution such that $x^i(0)=x^i(p)$ and $\sum_i[\dot{x}^i(t)]^2>0$ for sufficiently small t.

Then, the curve γ, defined parametrically by the functions $x^i(t)$, will be an integral curve of the field X that passes through the point p. On the basis of the theorem on the uniqueness of the solution of systems of differential equations, this curve is uniquely defined. Thus,

For $X_p\neq 0$, ***there exists exactly one integral curve*** $\gamma(t)$ ***of the field X such that*** $\gamma(0)=p$. (In this connection, we identify curves coinciding in a neighborhood of the point 0.)

We consider, in addition to curves, also smooth n-dimensional surfaces in M, that is, smooth mappings φ of a connected open set $G\subset R^n$ into the manifold M. As a rule, we shall assume the set G to be ***rectangular***, that is, the Cartesian product of n open intervals. For simplicity, we set $n=2$ although all that we are about to say will be applicable for arbitrary n. Thus, we shall consider surfaces of the form $\varphi: I^1\times I^2\to M$, where I^1 and I^2 are intervals of the real axis. For arbitrary fixed $t^1\in I^1$, each such surface $\varphi: I^1\times I^2\to M$

defines, in accordance with the formula

$$\varphi_{t^1}(t^2) = \varphi(t^1, t^2), \quad t^2 \in I^2,$$

a smooth curve $\varphi_{t^1}: I^2 \to M$. Analogously, for arbitrary fixed $t^2 \in I^2$, the surface φ defines, in accordance with the formula

$$\varphi_{t^2}(t^1) = \varphi(t^1, t^2), \quad t^1 \in I^1,$$

a smooth curve $\varphi_{t^2}: I^1 \to M$.

The two vectors $\frac{\partial}{\partial t^1}$ and $\frac{\partial}{\partial t^2}$, constituting the basis for the space $(I^1 \times I^2)_{(t^1, t^2)}$ are defined at every point (t^1, t^2) of the rectangle $I^1 \times I^2$. We shall denote the images of these vectors under the mapping

$$d\varphi_{(t^1, t^2)}: (I^1 \times I^2)_{(t^1, t^2)} \to M_{\varphi(t^1, t^2)}$$

by the symbols $\frac{\partial \varphi}{\partial t^1}(t^1, t^2)$ and $\frac{\partial \varphi}{\partial t^2}(t^1, t^2)$, respectively, and shall call them the ***basis tangent vectors*** of the surface φ at the point $p = \varphi(t^1, t^2)$. Obviously, they are the tangent vectors of the curves φ_{t^2} and φ_{t^1} at the points t^1 and t^2, respectively:

$$\frac{\partial \varphi}{\partial t^1}(t^1, t^2) = \dot{\varphi}_{t^2}(t^1), \quad \frac{\partial \varphi}{\partial t^2}(t^1, t^2) = \dot{\varphi}_{t^1}(t^2).$$

Suppose that to every point $(t^1, t^2) \in I^1 \times I^2$ is assigned a vector

$$X(t^1, t^2) \in M_{\varphi(t^1, t^2)}.$$

We shall call the mapping $(t^1, t^2) \to X(t^1, t^2)$ a vector field on the surface φ if, for an arbitrary function f that is smooth at the point $\varphi(t^1, t^2)$, the real function $g(t^1, t^2) = X(t^1, t^2) f$ is smooth at the point (t^1, t^2). Examples of vector fields on φ are the fields $\frac{\partial \varphi}{\partial t^1}(t^1, t^2)$ and $\frac{\partial \varphi}{\partial t^2}(t^1, t^2)$ consisting at each point of the basis tangent vectors constructed above.

If the surface φ ***is contained*** in a coordinate neighborhood U (that is, if $\varphi(I^1 \times I^2) \subset U$), then an arbitrary vector field $X(t^1, t^2)$ on φ is of the form

$$X(t^1, t^2) = X^i(t^1, t^2) \left(\frac{\partial}{\partial x^i} \right)_{\varphi(t^1, t^2)},$$

where $X^i(t^1, t^2)$ are smooth functions on $I^1 \times I^2$ (the ***components*** of the field $X(t^1, t^2)$).

For an arbitrary surface $\varphi: I^1 \times I^2 \to M$ and an arbitrary pair of functions $t^1(t)$ and $t^2(t)$ defined on some interval I and assuming values in the intervals I^1 and I^2, respectively, the formula

$$\gamma(t) = \varphi(t^1(t), t^2(t)), \quad t \in I,$$

defines a curve $\gamma(t)$ on M. We shall say that this curve belongs to the surface φ. It can be shown without difficulty that, just as in elementary calculus,

$$\dot{\gamma}(t) = \frac{\partial\varphi}{\partial t^1}(t^1(t),\ t^2(t))\,\dot{t}^1(t) + \frac{\partial\varphi}{\partial t^2}(t^1(t),\ t^2(t))\,\dot{t}^2(t).$$

The analogous formula for arbitrary n,

$$\dot{\gamma}(t) = \frac{\partial\varphi}{\partial t^i}(t^1(t),\ \ldots,\ t^n(t))\,\dot{t}^i(t),$$

where $\gamma(t)$ is the curve $\varphi(t^1(t), \ldots, t^n(t))$, is, of course, also valid.

The concept of tensor fields of arbitrary given type (r, s) on the surface φ can be defined as are vector fields, but we shall not need this concept.

13. The extension of tensor fields

We shall say that a tensor field (and, in particular, a vector field) $T(t)$ defined on a curve $\gamma(t)$ is ***extendable*** on a segment $\gamma|_C$ of the curve γ if there exists a tensor field T on M, known as the ***extension*** of the field $T(t)$, such that

$$T_{\gamma(t)} = T(t) \quad \text{for every} \quad t \in C.$$

In what follows, we shall say that a regular curve γ (resp. a segment $\gamma|_C$ of it) does not intersect itself if $\gamma(t_1) \neq \gamma(t_2)$ for $t_1 \neq t_2$, where t_1 and t_2 belong to I (resp. to C). We shall call a finite segment $\gamma|_C$ of a curve γ ***elementary*** if it does not intersect itself and if there exists on the manifold M a coordinate neighborhood U such that $\gamma(t) \in U$ for every $t \in C$.

It turns out that

Every tensor field T(t) on a regular curve $\gamma(t)$ is extendable on each of its elementary segments $\gamma|_C$.

In the first place, it is obvious that, to prove this assertion, it will be sufficient to construct the extension T of the field $T(t)$ to a coordinate neighborhood $U \supset \gamma(C)$. This is true because the finiteness of the segment C and the continuity of the mapping γ imply that the set $\gamma(C) \subset U$ is compact and hence possesses a neighborhood V with compact closure $\bar{V} \subset U$. Therefore, by virtue of the local epimorphism of the restriction mapping (cf. section 9), the field T defined on U coincides on V with some field defined on M, which will serve as the desired extension.

In the second place, to construct an extension of the field $T(t)$ onto U, it will be sufficient to assume that, for each of its components $T^{k_1 \ldots k_r}_{l_1 \ldots l_s}(t)$, there exists on M (or even on U) a smooth function $T^{k_1 \ldots k_r}_{l_1 \ldots l_s}$, such that

$$T^{k_1 \ldots k_r}_{l_1 \ldots l_s}(t) = T^{k_1 \ldots k_r}_{l_1 \ldots l_s}(\gamma(t)) \quad \text{for every} \quad t \in C.$$

This is true because the formula

$$T = T^{k_1 \ldots k_r}_{l_1 \ldots l_s} dx^{l_1} \otimes \ldots \otimes dx^{l_s} \otimes \frac{\partial}{\partial x^{k_1}} \otimes \ldots \otimes \frac{\partial}{\partial x^{k_r}}$$

will then define an extension T of the field $T(t)$ onto U. Thus, to prove the assertion made above, it will be sufficient to prove that

For an arbitrary function g(t) that is smooth at every point of a segment C, there exists on the manifold M a smooth function f such that

$$g(t) = f(\gamma(t)) \quad \textit{for every} \quad t \in C.$$

With this in mind, let us look at the functions

$$x^i(t) = x^i(\gamma(t))$$

that define the curve γ parametrically in a neighborhood U. Since $\dot{\gamma}(t_0) \neq 0$ for arbitrary $t_0 \in I$, there exists, by virtue of formula (3) of section 12, for every $t_0 \in C$ an index i_0 such that $\dot{x}^{i_0}(t_0) \neq 0$ and hence $\dot{x}^{i_0}(t) \neq 0$ in some neighborhood of the point t_0. Therefore, it follows from the familiar theorem in elementary calculus on inverse functions that there exists in some neighborhood of the point $x^{i_0}(t_0) \in R$ a smooth function y such that $y(x^{i_0}(t)) = t$ for arbitrary t sufficiently close to t_0, let us say, for $|t - t_0| < \delta_0$. Consider the function

$$\tilde{f}_{t_0}(p) = g(y(x^{i_0}(p))).$$

This function is defined in some neighborhood of the point $\gamma(t_0)$ and is smooth at $\gamma(t_0)$. Consequently, the point $\gamma(t_0)$ has a neighborhood $\tilde{U}_{t_0}$ in which the function $\tilde{f}_{t_0}$ coincides with some function f_{t_0} that is smooth on M. We can assume that $\gamma(t) \in \tilde{U}_{t_0}$ for $|t - t_0| < \delta_0$ and hence that

$$f_{t_0}(\gamma(t)) = \tilde{f}_{t_0}(\gamma(t)) = g(y(x^{i_0}(y(t)))) = g(t).$$

This shows us how to extend the function g "locally with respect to the parameter" (that is, for t close to t_0). However, the relation $\gamma(t) \in \tilde{U}_{t_0}$ may hold even for t not close to t_0 (the curve $\gamma(t)$ may have a "loop" located "close" to the point $\gamma(t_0)$). To be able to "glue" these local extensions, we must now shrink the neighborhood $\tilde{U}_{t_0}$ to some neighborhood U_{t_0} in such a way that, for $\gamma(t) \in U_{t_0}$, the number t will be close to the number t_0 (that is, so that there will be no "loop" of the curve $\gamma(t)$ in the reduced neighborhood U_{t_0}).

Since the segment C is finite, that is, compact, the set C' obtained by deleting from C points t such that $|t - t_0| < \delta_0$ is also compact. Consequently, its image $\gamma(C')$ under the mapping γ is closed. Also, it does not contain the point $\gamma(t_0)$ since, by hypothesis, the segment $\gamma|_C$ does not intersect itself. Therefore, the point $\gamma(t_0)$ possesses a neighborhood $U_{t_0} \subset \tilde{U}_{t_0}$ such that

$$U_{t_0} \cap \gamma(C') = \varnothing.$$

Now, it is clear that if $\gamma(t) \in U_{t_0}$ for $t \in C$, then $|t - t_0| < \delta_0$ and, therefore, in accordance with what was said above,

$$f_{t_0}(\gamma(t)) = g(t).$$

Now, suppose that W_{t_0} is an arbitrary neighborhood of the point $\gamma(t_0)$ the closure $\overline{W}_{t_0}$ of which is compact and contained in the neighborhood U_{t_0}. Since the set $\gamma(C)$ is compact, there exists a finite system of points $t_1, \ldots, t_r$ on the segment C such that the corresponding neighborhoods

$$W_1 = W_{t_1}, \ \ldots, \ W_r = W_{t_r}$$

constitute a covering of $\gamma(C)$. According to the lemma in section 3, there exists a smooth function h_i (for $i = 1, \ldots, r$) on the manifold M into the interval $[0, 1]$ that is equal to one on the compact set $\overline{W}_i$ and equal to 0 outside the open set $U_i = U_{t_i}$. Obviously, the function $h = h_1 + \ldots + h_r$ is nonzero on the set

$$W = W_1 \cup \ldots \cup W_r.$$

Therefore, the formula

$$\tilde{f} = \frac{f_1 h_1 + \ldots + f_r h_r}{h_1 + \ldots + h_r},$$

where $f_1, \ldots, f_r$ are functions f_{t_0} corresponding to points $t_1, \ldots, t_r$, defines a smooth function f on W. Since $\gamma(C) \subset W$, the number $\tilde{f}(\gamma(t))$ is defined for arbitrary $t \in C$. Since, by construction $f_i(\gamma(t)) = g(t)$ if $\gamma(t) \in U_i$ and $h_i(\gamma(t)) = 0$; otherwise we have

$$\tilde{f}(\gamma(t)) = g(t).$$

Finally, let V be an arbitrary neighborhood of the set $\gamma(C)$ the closure $\overline{V}$ of which is contained in the neighborhood W. Since the closure $\overline{V}$ is compact (it is contained in the compact set $\overline{W}$), there exists, by virtue of the local epimorphism of the restriction mapping (cf. section 3), a smooth function f on the manifold M that coincides with the function $\tilde{f}$ on the neighborhood V. Obviously, the function f possesses all the required properties. This completes the proof of the assertion made above.

Remark: An analogous assertion holds for any segment that does not intersect itself. We do not need this fact and therefore we leave it without proof.

To make clear the generality of the proposition just proven, it is worth noting that

For any regular curve $\gamma: I \to M$ *and any* $t_0 \in I$, *there exists an elementary segment* $\gamma|_C$ *of the curve* γ *such that* $t_0 \in C$.

Proof: Let $x^1, \ldots, x^m$ be an arbitrary system of local coordinates at the point $\gamma(t_0)$. As we have already seen, the assumed regularity of the curve γ implies that, among the functions $x^i(t)$ defining that curve parametrically, there exists a function $x^{i_0}(t)$ whose derivative $\dot{x}^{i_0}(t)$ is nonzero in some neighborhood of the point t_0 and hence on some closed interval C to which t_0 belongs. Since $\dot{x}^{i_0}(t) \neq 0$ on C, it follows that $x^{i_0}(t_1) \neq x^{i_0}(t_2)$ and hence $\gamma(t_1) \neq \gamma(t_2)$ for arbitrary distinct t_1 and t_2 in C. Consequently, the segment $\gamma|_C$ of the curve γ is elementary.

14. Submanifolds

Generalizing the concept of a regular curve, we shall say that a smooth mapping Φ of a smooth manifold N into a smooth manifold M is regular if, for an arbitrary point $p \in N$, the mapping

$$d\Phi_p : N_p \to M_{\Phi(p)}$$

is monomorphic (that is, has a trivial kernel).

We shall say that a manifold N is a submanifold of a manifold M if $N \in M$ and the embedding mapping $\iota: N \to M$ is smooth and regular. In this case, we shall say that those vectors in M_p (where $p \in N$) that are images of vectors in N_p under the mapping $d\iota_p$, are tangent to the submanifold N at the point p. As a rule, we shall identify with the space N_p the subspace $d\iota_p(N_p)$ of vectors in M_p that are tangent to the submanifold N; that is, we shall consider the monomorphic mapping $d\iota_p$ as an embedding.

In the special case in which $N = M$, we obtain the result that, in accordance with this definition, an arbitrary vector in M_p is tangent to the manifold M at the point p. This explains the name "tangent space" for the space M_p.

Let p denote an arbitrary point of a submanifold N of a manifold M, let $x^1, \ldots, x^m$ be an arbitrary system of local coordinates on the manifold M at the point p, and let $y^1, \ldots, y^n$ be an arbitrary system of local coordinates on the submanifold N at the point p. Consider the functions

$$x^1|_N = x^1 \circ \iota, \ \ldots, \ x^m|_N = x^m \circ \iota,$$

where ι is the embedding mapping $N \to M$. These functions are smooth at the point p. Therefore, the relations

$$x^1 \circ \iota = u^1(y^1, \ldots, y^n), \ \ldots, \ x^m \circ \iota = u^m(y^1, \ldots, y^n),$$

where $u^1, \ldots, u^m$ are smooth functions of n real variables, are valid close to that point on the submanifold N. By differentiating these relations, we easily see that

$$\left(\frac{\partial}{\partial y^i}\right)_p = \left(\frac{\partial (x^j \circ \iota)}{\partial y^i}\right)_p \left(\frac{\partial}{\partial x^j}\right)_p, \quad i = 1, \ldots, n.$$

(In accordance with what was said above, we are now considering the vectors $\left(\frac{\partial}{\partial y^i}\right)_p \in N_p$ as vectors in the space M_p.) Since the vectors $\left(\frac{\partial}{\partial y^i}\right)_p$, for $i = 1, \ldots, n$, are linearly independent, it follows that the matrix

$$\left(\left(\frac{\partial (x^j \circ \iota)}{\partial y^i}\right)_p\right)$$

is of rank n and therefore it possesses a nonzero minor of order n. Suppose, for definiteness, that

$$\det \left| \frac{\partial (x^j \circ \iota)}{\partial y^i} \right| \neq 0, \quad i, j = 1, \ldots, n.$$

Then, in accordance with the theorem on change of local coordinates, the functions $x^1 \circ \iota, \ldots, x^n \circ \iota$ constitute a system of local coordinates at the point p on the submanifold N. Thus, we have proved that

For an arbitrary system of local coordinates $x^1, \ldots, x^m$ *on a manifold M at a point* $p \in N$, *among the functions*

$$x^1|_N = x^1 \circ \iota, \ldots, x^m|_N = x^m \circ \iota$$

there exist n functions constituting a system of local coordinates on the submanifold N.

It is easy to see that

An arbitrary open submanifold W of a manifold M is a submanifold of it the dimension of which is equal to the dimension of the manifold M.

This is true because, obviously, $W_p = M_p$ for an arbitrary point $p \in W$.

It turns out that the converse is also true:

An arbitrary n-dimensional submanifold N of a manifold M is an open submanifold.

Proof: In accordance with what was proven above, for an arbitrary system of local coordinates $x^1, \ldots, x^n$ on a manifold M, at a point $p \in N$ the functions $x^1|_N, \ldots x^n|_N$ constitute a system of local coordinates on the submanifold N. Consequently, the corresponding coordinate neighborhood of the point p in the submanifold N is open in the manifold M; that is, the point p is an interior point of the set N (with respect to the manifold M). But this means that N is open in M.

We obtain an important example by considering an arbitrary smooth function φ and a smooth manifold M. Let a denote an arbitrary real number and let $[\varphi = a]$ denote the corresponding ***level surface*** of the function φ, that is, the set of all points $p \in M$ at which $\varphi(p) = a$ (we are assuming that this set is not empty). We shall say that the level surface $[\varphi = a]$ is regular if $d\varphi_p \neq 0$ for every point $p \in [\varphi = a]$.

Every level surface $[\varphi = a]$, being a subset of a smooth manifold M, is naturally a smooth premanifold (cf. section 1). It turns out that

If a level surface $[\varphi = a]$ *is regular, it is an* $(m-1)$*-dimensional smooth manifold and a submanifold of the manifold* M.

Proof: Let p denote an arbitrary point of the level surface $[\varphi = a]$ and let $x^1, \ldots, x^m$ be an arbitrary system of local coordinates of the manifold M at the point p. Because of the regularity of the level surface $[\varphi = a]$, at least one partial derivative $\partial\varphi/\partial x^i$ of the function φ is nonzero at the point p. Suppose for definiteness that

$$\left(\frac{\partial \varphi}{\partial x^1}\right)_p \neq 0.$$

Then, by virtue of the theorem on change of local coordinates (cf. section 2), the functions $\varphi, x^2, \ldots, x^m$ will also be local coordinates of the manifold M at the point p. Suppose that U is the corresponding coordinate neighborhood and that $\xi : U \to R^m$ is the corresponding coordinate homeomorphism

Let p be a point on the level surface $[\varphi = a]$. Consider the functions

$$y^1 = x^2|_V, \ \ldots, \ y^{m-1} = x^m|_V$$

in the neighborhood $V = U \cap [\varphi = a]$ of p. By definition, these functions are smooth at an arbitrary point $q \in V$. Furthermore, if a function f on $[\varphi = a]$ is smooth at the point $q \in V$ and is the restriction to U of some function

$$g = u(\varphi, \ x^2, \ \ldots, \ x^m)$$

that is smooth at q, then

$$f = u(a, \ y^1, \ \ldots, \ y^{m-1}),$$

in some neighborhood of the point q; that is, f depends smoothly on the functions $y^1, \ldots, y^{m-1}$ close to the point q. Finally, the mapping $\eta : V \to R^{m-1}$ defined by the formula

$$\eta(q) = (y^1(q), \ \ldots, \ y^{m-1}(q)), \quad q \in V,$$

is connected with the coordinate homeomorphism $\xi : U \to R^m$ by the formula

$$\xi(q) = (a, \ \eta(q)), \quad q \in V.$$

From this it follows that this mapping maps the neighborhood V homeomorphically onto the intersection of the set ξU and the hyperplane $t^1 = a$. This proves that the functions $y^1, \ldots, y^{m-1}$ are local coordinates of the premanifold $[\varphi = a]$ in the neighborhood V of the point p. Since the point p is arbitrary, this proves that the premanifold $[\varphi = a]$ is indeed a smooth manifold.

It remains to show that this manifold is a submanifold of the manifold M. But this is almost obvious. Let $\iota : [\varphi = a] \to M$ be the embedding mapping. Then, by definition, $x^{k+1} \circ \iota = y^k$, for $k = 1, \ldots, m-1$. Therefore,

$$d\iota_p \left(\frac{\partial}{\partial y^k} \right) = \frac{\partial}{\partial x^{k+1}}$$

(cf. section 6), from which it follows immediately that, for an arbitrary point $p \in [\varphi = a]$, the mapping $d\iota_p$ is monomorphic. This completes the proof of the assertion made above.

At every point $p \in [\varphi = a]$, the space $[\varphi = a]_p$ tangent to the submanifold $[\varphi = a]$ is an $(m-1)$-dimensional subspace of the space M_p. Since $\varphi \circ \iota =$ const on $[\varphi = a]$, we have $d(\varphi \circ \iota)_p = 0$. On the other hand, $d(\varphi \circ \iota)_p = d\varphi_p \circ d\iota_p$. Therefore, $d\varphi_p(A) = 0$ for an arbitrary vector $A \in [\varphi = a]_p$. Since the covector $d\varphi_p$ is nonzero, the vectors $A \in M_p$ such that $d\varphi_p(A) = 0$ constitute an $(m-1)$-dimensional subspace of M_p. Since the space $[\varphi = a]_p$ is also $(m-1)$-dimensional, it follows that

The vector $A \in M_p$ belongs to the subspace $[\varphi = a]_p$ (that is, it is tangent to the submanifold $[\varphi = a]$) if and only if

$$d\varphi_p(A) = 0.$$

An example of a function φ illustrating the assertion proven above is the function

$$\varphi = (t^1)^2 + \ldots + (t^m)^2$$

on the space R^m. The corresponding level surface $[\varphi = a]$ is the sphere of radius $\sqrt{a}$. Since

$$d\varphi = 2t^1\, dt^1 + \ldots + 2t^m\, dt^m$$

this level surface is regular for all $a > 0$. Thus,

The spheres in the space R^m are submanifolds.

A nonregular level surface $[\varphi = a]$ is not in general a smooth manifold. However, if we delete from it all points p for which $d\varphi_p = 0$ (such points are called *critical* points of the function φ), the remaining set $[\varphi = a]'$, which is open in $[\varphi = a]$ will obviously be a smooth manifold and a submanifold of the manifold M.

An arbitrary $(m-1)$-dimensional submanifold N of a manifold M cannot in general be represented in the form of a level surface of a function φ that is smooth on M. This can be done only locally; that is,

For an arbitrary point p of an $(m-1)$-dimensional submanifold N of a manifold M, there exists a smooth function φ on M such that, close to the point p (that is, in some neighborhood of p), the submanifold N coincides with the level surface $[\varphi = 0]$.

This assertion is a special case of the following more general proposition, which applies to an arbitrary n-dimensional (with $0 \leqslant n \leqslant m$) submanifold N of a manifold M:

For an arbitrary point $p \in N$ on a manifold M, there exists a system of local coordinates $x^1, \ldots, x^m$ at the point p such that a necessary and sufficient condition for the point q of the corresponding coordinate neighborhood U to belong to the submanifold N is

$$x^{n+1}(q) = 0, \ \ldots, \ x^m(q) = 0.$$

In this case, we shall say that the submanifold N is defined in the neighborhood U by the equations

$$x^{n+1} = 0, \ \ldots, \ x^m = 0.$$

To prove this important proposition, let us look at a basis $A_1, \ldots, A_m$ of the space M_p such that the first n vectors $A_1, \ldots, A_n$ in it constitute a basis for the subspace N_p. As we know (cf. section 6) there exists at the point p on the manifold M a system of local coordinates $x^1, \ldots, x^m$ such that

$$\left(\frac{\partial}{\partial x^1}\right)_p = A_1, \ \ldots, \ \left(\frac{\partial}{\partial x^m}\right)_p = A_m.$$

Furthermore, we know that from the functions

$$y^1 = x^1|_N, \ \ldots, \ y^m = x^m|_N$$

we can choose n functions constituting a system of local coordinates at the point p of the submanifold N. Since by hypothesis, the vectors $\left(\frac{\partial}{\partial x^1}\right)_p, \ldots, \left(\frac{\partial}{\partial x^n}\right)_p$ constitute a basis for the subspace N_p, we can obviously take for these functions the first n functions $y^1, \ldots, y^n$. But then, for $j = n+1, \ldots, m$ and near p the function y^j depends smoothly on the functions $y^1, \ldots, y^n$; that is, there exists a smooth function $u^j(t^1, \ldots, t^n)$ of the real variables $t^1, \ldots, t^n$ such that, near p on the submanifold N,

$$y^j = u^j(y^1, \ldots, y^n), \quad j = n+1, \ldots, m.$$

Let us set

$$\begin{aligned} \bar{x}^1 &= x^1, \\ &\vdots \\ \bar{x}^n &= x^n, \\ \bar{x}^{n+1} &= x^{n+1} - u^{n+1}(x^1, \ldots, x^n), \\ &\vdots \\ \bar{x}^m &= x^m - u^m(x^1, \ldots, x^n). \end{aligned}$$

According to the theorem on change of local coordinates, the functions $\bar{x}^1, \ldots, \bar{x}^m$ in some neighborhood $U \subset M$ of the point p are local coordinates on M. On the other hand, it is easy to see that the point $q \in U$ belongs to the submanifold N if and only if $\bar{x}^j(q) = 0$ for $j = n+1, \ldots m$. . If we again denote the coordinates $\bar{x}^i$ by the symbols x^i, we see that the proposition formulated above is completely proven.

We note that this system of local coordinates obviously has the following property:

The functions

$$x^1|_N, \ldots, x^n|_N$$

constitute a system of local coordinates on the submanifold N.

Now, let N' denote an n'-dimensional submanifold of the manifold M and let $p \in N$. Suppose that its tangent space N'_p intersects the space N_p only at zero. If we make the transformation of coordinates constructed above for the manifold N', we easily obtain the result:

The system of local coordinates $x^1, \ldots, x^m$ *referred to in the preceding proposition can be chosen so that the submanifold N′ near p is defined by the equations*

$$x^1 = 0, \ldots, x^n = 0, \; x^{n+n'+1} = 0, \ldots, x^m = 0.$$

In particular, let $\gamma(t)$ denote an arbitrary regular curve that passes through the point p and nontangentially meets the submanifold N (that is, such that its tangent vector $\dot{\gamma}_p$ at that point does not belong to the space N_p). Since every regular curve that does not intersect itself can be regarded in a natural way as a one-dimensional submanifold, it follows from the assertion just proven that

At a point p on a manifold M, there exists a system of local coordinates $x^1, \ldots, x^m$ *such that, close to the point p, the submanifold N is defined by the equations* $x^{n+1}=0, \ldots, x^m=0$ *and the curve* γ *by the equations* $x^1 = 0, \ldots, x^{m-1} = 0$.

15. Products of manifolds

Let M and N denote arbitrary smooth manifolds of dimensions m and n, respectively, and let $M \times N$ denote the Cartesian product of these manifolds, treated as topological spaces. Every function g on M defines a function $\hat{g}$ on $M \times N$ by the formula

$$\hat{g}(p, q) = g(p), \quad p \in M, \; q \in N.$$

In other words,

$$\hat{g} = g \circ \varkappa_1,$$

where $\varkappa_1 : M \times N \to M$ is the natural projection. Analogously, every function h on N defines a function $\hat{h}$ on $M \times N$ by the formula

$$\hat{h} = h \circ \varkappa_2,$$

where $\varkappa_2$ is the natural projection

$$M \times N \to N.$$

We shall say that a function f on $M \times N$ is smooth on $M \times N$ if, close to an arbitrary point $(p, q) \in M \times N$, it depends smoothly on functions of the type $\hat{g}$ and $\hat{h}$, where $g \in \mathfrak{G}(M)$ and $h \in \mathfrak{G}(N)$. It turns out that

By virtue of this definition, the space $M \times N$ becomes a smooth manifold of dimension $m+n$.

That this definition makes the space $M \times N$ a smooth premanifold is obvious. On the other hand, it is obvious that, for arbitrary local coordinates $x^1, \ldots, x^m$ and $y^1, \ldots, y^n$ at points p and q of the manifolds M and N, respectively, the functions

$$\hat{x}^1, \ldots, \hat{x}^m, \hat{y}^1, \ldots, \hat{y}^n$$

constitute a system of local coordinates at the point $(p, q) \in M \times N$.

We shall call this smooth manifold $M \times N$ the product of the manifolds M and N. Obviously,

The natural projections

$$\varkappa_1 : M \times N \to M \text{ and } \varkappa_2 : M \times N \to N$$

are smooth mappings.

The product $M_1 \times \ldots \times M_r$ of an arbitrary finite number of smooth manifolds $M_1, \ldots, M_r$ is defined analogously. Functions that depend smoothly, near an arbitrary point in this product, on functions of the form $g_i \circ \varkappa_i$, where

$$\varkappa_i : M_1 \times \ldots \times M_r \to M_i$$

is the natural projection and g_i is an arbitrary function that is smooth on M_i, are smooth functions on this product. Then, the natural projections $\varkappa_i$ are also smooth mappings.

An arbitrary vector A at a point (p, q) of the manifold $M \times N$ defines vectors A_1 and A_2 at points p and q of the manifolds M and N, respectively, by

$$A_1 = (d\varkappa_1)_{(p,q)}(A), \quad A_2 = (d\varkappa_2)_{(p,q)}(A).$$

Obviously, for an arbitrary system of local coordinates $x^1, \ldots, x^m$ at a point q of the manifold M,

$$A\hat{x}^i = A_1 x^i, \quad i = 1, \ldots, m.$$

Analogously, for an arbitrary system of local coordinates $y^1, \ldots, y^m$ at a point p of the manifold N,

$$A\hat{y}^j = A_2 y^j, \quad j = 1, \ldots, n.$$

This tells us, first, that $A = 0$ if and only if $A_1 = A_2 = 0$ and, second, that, for arbitrary vectors $A_1 \in M_p$ and $A_2 \in N_q$, there exists a vector $A \in (M \times N)_{(p,q)}$ to which these vectors correspond. In other words,

The mappings

$$(d\varkappa_1)_{(p,q)} : (M \times N)_{(p,q)} \to M_p, \ (dx_2)_{(p,q)}: (M \times N)_{(p,q)} \to N_q$$

define a decomposition of the space $(M \times N)_{(p,q)}$ *into the direct sum of the spaces* M_p *and* N_q*.*

Here, it is clear that this decomposition is compatible with the operations of the vectors A and functions f that are smooth at the point (p, q); that is, for any such function,

$$Af = A_1 f + A_2 f,$$

where by $A_1 f$, for example, we mean the result obtained by applying the vector A_1 to the function f, treated as a function of a single point p (with the point q fixed).

In what follows, we shall sometimes find it convenient to denote the vector A by the symbol (A_1, A_2).

Obviously, an analogous decomposition into a direct sum is possible for products of an arbitrary finite number of manifolds.

CHAPTER 2

SPACES OF AFFINE CONNECTION

1. Affine connections

Let M denote an arbitrary smooth manifold. A function ∇ that assigns to each vector field $X \in \mathfrak{a}^1(M)$ a linear (but not necessarily $\mathfrak{a}(M)$-linear) mapping

$$\nabla_X : \mathfrak{a}^1(M) \to \mathfrak{a}^1(M)$$

is called an *affine connection* on the manifold M if, for arbitrary functions $f, g \in \mathfrak{a}(M)$ and arbitrary fields $X, Y \in \mathfrak{a}^1(M)$,

$$\nabla_{fX+gY} = f\nabla_X + g\nabla_Y \qquad (\mathfrak{a}(M)\text{-linearity in } X),$$

$$\nabla_X(fY) = Xf \cdot Y + f \cdot \nabla_X Y \qquad \text{(the rule for differentiation of a product).}$$

The manifold M on which an affine connection ∇ is defined is called the space of the affine connection and the mapping ∇_X is called *covariant differentiation* along the vector field X (relative to the given affine connection ∇).

The concept of an affine connection has a local character; that is,

If at least one of the vector fields X and Y vanishes on an open submanifold W of a space M of affine connection, then the vector field $\nabla_X Y$ also vanishes on W.

Proof: Suppose that f is a smooth function on M that is equal to unity at some point $p \in W$ and equal to 0 outside W. If $Y = 0$ on W, then $fY = 0$ on M and hence $\nabla_X(fY) = 0$ on M. But

$$\nabla_X(fY) = Xf \cdot Y + f \cdot \nabla_X Y,$$

so that

$$f \cdot \nabla_X Y = -Xf \cdot Y \quad \text{on } M$$

and, therefore,

$$(\nabla_X Y)_p = -(Xf)(p) \cdot Y_p = 0,$$

since $f(p) = 1$ and $Y_p = 0$.

Analogously, if $X = 0$ on W, then $fX = 0$ on M and hence $\nabla_{fX}Y = 0$ on M. But

$$\nabla_{fX}Y = f\nabla_X Y, \quad \text{so that} \quad (\nabla_X Y)_p = 0,$$

since $f(p) = 1$. Since p is an arbitrary point of the set W, this proves that $\nabla_X Y = 0$ on W in both cases.

Now, let X and Y denote arbitrary vector fields on an open submanifold W of a space of affine connection M. As we know, for an arbitrary point $p \in W$, there exists a neighborhood V of p contained in W and vector fields X' and Y' on M such that $X' = X$ and $Y' = Y$ on V. Let us define an affine connection $\nabla|_W$ on W by setting

$$((\nabla|_W)_X Y)_p = (\nabla_{X'} Y')_p, \quad p \in W.$$

Because of the local nature of an affine connection, the vector $(\nabla_{X'} Y')_p$, for an arbitrary point $p \in W$, is independent of the choice of fields X' and Y'. Therefore, this formula determines unambiguously a vector field $(\nabla|_W)_X Y$ on W. That the function $\nabla|_W$ we have constructed in an affine connection can be verified directly.

We shall call the affine connection $\nabla|_W$ a restriction of the connection ∇ to W and we shall call the open submanifold W equipped with this connection an open subspace of the space M. When there is no danger of confusion, we shall denote the affine connection $\nabla|_W$ on W with the same symbol ∇ as that used for the affine connection on the entire space M.

In particular, the affine connection $\nabla|_U$ is defined on every coordinate neighborhood U of the space M. Let $X_1, \ldots, X_m$ denote an arbitrary basis for the module $\mathfrak{a}^1(U)$ and let

$$X = X^i X_i, \quad Y = Y^j X_j$$

denote arbitrary vector fields on U. Then, as one can easily see,

$$\nabla_X Y = X^i \cdot X_i Y^j \cdot X_j + X^i Y^j \nabla_{X_i}(X_j) \quad \text{on} \quad U.$$

If we set

$$\nabla_{X_i}(X_j) = \Gamma_{ij}^k X_k,$$

where the Γ_{ij}^k are smooth functions on U (the so-called coefficients of the connection ∇ relative to the basis $X_1, \ldots, X_m$), we obtain the following expression for the components $(\nabla_X Y)^k$ of the field $\nabla_X Y$:

$$(\nabla_X Y)^k = (X_i Y^k + \Gamma_{ij}^k Y^j) X^i. \tag{1}$$

In particular, if $X_i = \partial/\partial x^i$, then

$$(\nabla_X Y)^k = \left(\frac{\partial Y^k}{\partial x^i} + \Gamma_{ij}^k Y^j\right) X^i. \tag{1'}$$

2. The Curvature and Torsion Tensors

Let X and Y denote arbitrary vector fields defined on a space M of affine connection. We define

$$K(X, Y) = \nabla_X Y - \nabla_Y X - [X, Y],$$
$$R(X, Y) = \nabla_X \nabla_Y - \nabla_Y \nabla_X - \nabla_{[X, Y]}.$$

Thus, $K(X, Y)$ is a vector field on M, and $R(X, Y)$ is a mapping $\mathfrak{G}^1(M) \to \mathfrak{G}^1(M)$. Obviously, the following "antisymmetry properties" hold:

$$K(X, Y) = -K(Y, X), \quad R(X, Y) = -R(Y, X).$$

It can also be proven directly that, for any three fields X, Y, $Z \in \mathfrak{G}^1(M)$,

$$R(X, Y)Z + R(Y, Z)X + R(Z, X)Y =$$
$$= K(X, [Y, Z]) + K(Y, [Z, X]) + K(Z, [X, Y]) +$$
$$+ \nabla_X K(Y, Z) + \nabla_Y K(Z, X) + \nabla_Z K(X, Y).$$

In particular,

If $K = 0$, then

$$R(X, Y)Z + R(Y, Z)X + R(Z, X)Y = 0.$$

This last identity is known as ***Bianchi's identity.***

We shall usually find it more convenient to consider not the field $K(X, Y)$ and the mapping $R(X, Y)$ but the tensor fields K and R defined by

$$K(\omega; X, Y) = \omega(K(X, Y)), \quad \omega \in \mathfrak{G}_1(M), \quad X, Y \in \mathfrak{G}^1(M),$$
$$R(\omega; Z, X, Y) = \omega(R(X, Y)Z), \quad \omega \in \mathfrak{G}_1(M), \quad X, Y, Z \in \mathfrak{G}^1(M).$$

It can be shown by direct verification that these formulas do indeed define tensor fields. We shall call the field K [of type (1, 2)] the ***torsion tensor field*** and we shall call the field R [of type 1, 3)] the ***curvature tensor field*** of the space of affine connection M in question.

At every point $p \in M$, the field K defines a tensor $K_p \in \mathfrak{G}_2^1(p)$, called the ***torsion tensor*** at the point p. For arbitrary vectors A, $B \in \mathfrak{G}^1(p)$, the function

$$K_p(A, B)(\alpha) = K_p(\alpha; A, B)$$

of the covector $\alpha \in \mathfrak{G}_1(p)$ is a tensor of type (1, 0), that is, a vector. Thus, the tensor K_p enables us to assign to arbitrary vectors A, $B \in M_p$ a vector $K_p(A, B) \in M_p$.

Analogously, at every point $p \in M$, the field R defines a tensor $R_p \in \mathfrak{G}_3^1(p)$, known as the ***curvature tensor*** at the point p. For arbitrary vectors $A, B, C \in \mathfrak{G}^1(p)$, the function

$$[R_p(A, B)C](\alpha) = R_p(\alpha; C, A, B)$$

of the covector $\alpha \in \mathfrak{6}_1(p)$ is a tensor of type $(1, 0)$, that is, a vector. Obviously, the mapping $C \to R_p(A, B)C$ is linear. Thus, the tensor R_p enables us to assign to arbitrary vectors $A, B \in M_p$ a linear transformation $R_p(A, B)$ of the space M_p.

Now, let U denote an arbitrary coordinate neighborhood in the space M and let $X_1, \ldots, X_m$ denote an arbitrary basis for the $\mathfrak{6}(U)$-module $\mathfrak{6}^1(U)$. By setting

$$[X_i, X_j] = c_{ij}^k X_k,$$

and using formula (1) of section 1, we can easily show that, for arbitrary vector fields $X, Y \in \mathfrak{6}^1(U)$, the components $K(X, Y)^k$ of the vector field $K(X, Y)$ on U are given by the formula

$$K(X, Y)^k = (\Gamma_{ij}^k - \Gamma_{ji}^k - c_{ij}^k) X^i Y^j.$$

From this, we obtain for the components K_{ij}^k of the tensor field K the formula

$$K_{ij}^k = \Gamma_{ij}^k - \Gamma_{ji}^k - c_{ij}^k. \tag{1}$$

In particular, if the basis $X_1, \ldots, X_m$ is holonomic, then

$$K_{ij}^k = \Gamma_{ij}^k - \Gamma_{ji}^k. \tag{1'}$$

Consequently,

The field K of the torsion tensors is equal to 0 if and only if, for an arbitrary system of local coordinates on the space M, the corresponding components of the connection Γ_{ij}^k are symmetric with respect to their subscripts:

$$\Gamma_{ij}^k = \Gamma_{ji}^k.$$

Analogously, if we use formula (1) of section 1 to evaluate the components of the fields $\nabla_X \nabla_Y Z$ and $\nabla_Y \nabla_X Z$, we can easily show that, for any three fields $X, Y, Z \in \mathfrak{6}^1(U)$, the field $R(X, Y)Z \in \mathfrak{6}^1(U)$ has components

$$(R(X, Y)Z)^k =$$
$$= (\Gamma_{is}^k \Gamma_{jl}^s - \Gamma_{js}^k \Gamma_{il}^s + X_i \Gamma_{jl}^k - X_j \Gamma_{il}^k - c_{ij}^s \Gamma_{sl}^k) X^i Y^j Z^l,$$

that is, that the components R_{lij}^k of the tensor field R are given by

$$R_{lij}^k = \Gamma_{is}^k \Gamma_{jl}^s - \Gamma_{js}^k \Gamma_{il}^s + X_i \Gamma_{jl}^k - X_j \Gamma_{il}^k - c_{ij}^s \Gamma_{sl}^k. \tag{2}$$

In particular, if the basis $X_1, \ldots, X_m$ is holonomic, then

$$R_{lij}^k = \Gamma_{is}^k \Gamma_{jl}^s - \Gamma_{js}^k \Gamma_{il}^s + \frac{\partial \Gamma_{jl}^k}{\partial x^i} - \frac{\partial \Gamma_{il}^k}{\partial x^j}. \tag{2'}$$

3. Covariant Differentiation Along a Curve

Let $\gamma: I \to M$ denote an arbitrary regular curve in a space M of affine connection and let $X(t)$ denote a vector field on the curve γ. Consider an arbitrary elementary segment $\gamma|_C$ of the curve γ. As we know (cf. section 13 of Chapter I), there exists a vector field X on the manifold M such that

$$X_{\gamma(t)} = X(t) \quad \text{for every} \quad t \in C.$$

Similarly, there exists a vector field A on M such that

$$A_{\gamma(t)} = \dot{\gamma}(t) \quad \text{for every} \quad t \in C.$$

Let us define a vector field $\frac{\nabla X}{dt}(t)$ on $\gamma|_C$ by

$$\frac{\nabla X}{dt}(t) = (\nabla_A X)_{\gamma(t)} \quad \text{for every} \quad t \in C.$$

We have

The field $\frac{\nabla X}{dt}(t)$ *is independent of the choice of the fields* X *and* A; *that is, it is determined exclusively by the field* $X(t)$ *and the curve* γ.

To prove this assertion, let us consider local coordinates $x^1, \ldots, x^m$ defined in a coordinate neighborhood $U \supset \gamma(C)$. Suppose that

$$X(t) = X^i(t)\left(\frac{\partial}{\partial x^i}\right)_{\gamma(t)} \quad \text{on } C$$

and

$$X = X^i \frac{\partial}{\partial x^i} \quad \text{on } U,$$

where the $X^i = Xx^i$ are smooth functions on U and the $X^i(t)$ are smooth functions on C connected with the functions X^i by

$$X^i(t) = X^i(\gamma(t)).$$

Analogously, suppose that

$$\dot{\gamma}(t) = A^i(t)\left(\frac{\partial}{\partial x^i}\right)_{\gamma(t)} \quad \text{on } C$$

and

$$A = A^i \frac{\partial}{\partial x^i} \quad \text{on } U,$$

where $A^i = Ax^i$ and $A^i(t) = A^i(\gamma(t))$. Then,

$$(\nabla_A X)^k = A^i \frac{\partial X^k}{\partial x^i} + \Gamma^k_{ij} A^i X^j, \qquad k = 1, \ldots, m,$$

where the Γ_{ij}^k are the coefficients of the connection ∇. If we denote the components of the vector

$$\frac{\nabla X}{dt}(t) = (\nabla_{\mathrm{A}} X)_{\gamma(t)}$$

relative to the basis

$$\left(\frac{\partial}{\partial x_1}\right)_{\gamma(t)}, \ldots, \left(\frac{\partial}{\partial x^m}\right)_{\gamma(t)}$$

by the symbols $\frac{\nabla X^k}{dt}(t)$, for $k = 1, \ldots, m$, we immediately see that

$$\frac{\nabla X^k}{dt}(t) = \mathrm{A}^i(t)\frac{\partial X^k}{\partial x^i}(\gamma(t)) + \Gamma_{ij}^k(t)\,\mathrm{A}^i(t)\,X^j(t),$$

where

$$\Gamma_{ij}^k(t) = \Gamma_{ij}^k(\gamma(t)).$$

But we know (see formula (3), section 12, Chapter I) that

$$\mathrm{A}^i(t) = \dot{x}^i(t).$$

Therefore, in accordance with the rule for differentiating a composite function, we immediately get

$$\mathrm{A}^i(t)\frac{\partial X^k}{\partial x^i}(\gamma(t)) = \frac{dX^k(t)}{dt}.$$

Thus,

$$\frac{\nabla X^k}{dt}(t) = \frac{dX^k(t)}{dt} + \Gamma_{ij}^k(t)\,\dot{x}^i(t)\,X^j(t), \qquad k = 1, \ldots, m. \tag{1}$$

The right-hand side of this formula is independent of the choice of the fields X and A. This completes the proof of the assertion made above.

Since an arbitrary point of a regular curve γ belongs to some elementary segment, the construction described above defines the vector $\frac{\nabla X}{dt}(t)$ unambiguously for arbitrary $t \in I$. Obviously, the mapping $t \to \frac{\nabla X}{dt}(t)$ for $t \in I$ satisfies the smoothness condition; that is, the vectors $\frac{\nabla X}{dt}(t)$ constitute a vector field on the curve $\gamma(t)$. We shall say that this field is obtained by ***covariant differentiation of the field*** $X(t)$ ***along the curve*** γ.

Obviously, covariant differentiation ∇/dt on the set of all fields $X(t)$ is linear; that is, for any two fields $X_1(t)$ and $X_2(t)$ on the curve $\gamma(t)$,

$$\frac{\nabla}{dt}(X_1(t)+X_2(t))=\frac{\nabla X_1}{dt}(t)+\frac{\nabla X_2}{dt}(t).$$

Furthermore, when we multiply a vector field by a real function, covariant differentiation obeys the rule for differentiation of a product; that is, for an arbitrary function $f(t)$, where $t\in I$, and an arbitrary field $X(t)$, on the curve γ, we have

$$\frac{\nabla}{dt}(f(t)X(t))=\frac{df(t)}{dt}X(t)+f(t)\frac{\nabla X}{dt}(t).$$

In an arbitrary system of local coordinates $x^1, \ldots, x^m$, the components $\frac{\nabla X^k}{dt}(t)$ of the field $\frac{\nabla X}{dt}(t)$ relative to the basis

$$\left(\frac{\partial}{\partial x^1}\right)_{\gamma(t)}, \ldots, \left(\frac{\partial}{\partial x^m}\right)_{\gamma(t)}$$

are obviously expressed by formulas (1) independently of whether the segment of the curve γ lying in the corresponding coordinate neighborhood is elementary or not.

Furthermore, formulas (1) can be obtained on the basis of a "coordinate" definition of the operation ∇/dt since, as we can easily verify, the numbers $\frac{\nabla X^k}{dt}(t)$ given by these formulas serve as coordinates of a completely defined vector (that is, defined independently of the choice of local coordinates $x^1, \ldots, x^m$). It is important to note that this new definition can also be applied in the case in which the curve $\gamma(t)$ is nonregular. Thus,

The operation ∇/dt of covariant differentiation of vector fields along a curve is defined for an arbitrary smooth curve $\gamma(t)$.

Keeping this in mind, let us consider, in the space M, an arbitrary smooth surface $\varphi: I^1\times I^2\to M$ and a vector field $X(t^1, t^2)$ on the surface φ. (As in section 12 of Chapter I, we shall assume that the surface φ is 2-dimensional although what follows is applicable with appropriate modifications for arbitrary dimensions.) For arbitrary fixed $t^2\in I^2$, the vectors $X(t^1, t^2)$ constitute a field $X_{t^2}(t^1)$ on a smooth curve $\varphi_{t^2}(t^1)=\varphi(t^1, t^2)$. Therefore, the field $\nabla X_{t^2}/dt^1$ is defined on that curve. Obviously, the formula

$$\frac{\nabla X}{\partial t^1}(t^1, t^2)=\frac{\nabla X_{t^2}}{dt^1}(t^1)$$

defines a vector $\frac{\nabla X}{\partial t^1}(t^1, t^2)$ on the surface φ. Analogously, the formula

$$\frac{\nabla X}{\partial t^2}(t^1, t^2)=\frac{\nabla X_{t^1}}{dt^2}(t^2),$$

where $X_{t^1}(t^2) = X(t^1, t^2)$, defines on φ a vector field $\frac{\nabla X}{\partial t^2}(t^1, t^2)$. We shall call the operations $\nabla/\partial t^1$ and $\nabla/\partial t^2$ the operations of *covariant differentiation on the surface* φ *with respect to the parameters* t^1 *and* t^2, *respectively.*

Let $x^1, \ldots, x^m$ denote local coordinates at a point $p = \varphi(t^1, t^2)$. Obviously, the components $\frac{\nabla X^k}{\partial t^1}(t^1, t^2)$ of the vector $\frac{\nabla X}{\partial t^1}(t^1, t^2)$ relative to the basis $\left(\frac{\partial}{\partial x^1}\right)_p, \ldots, \left(\frac{\partial}{\partial x^m}\right)_p$ are expressed by the formula

$$\frac{\nabla X^k}{\partial t^1}(t^1, t^2) = \frac{\partial X^k(t^1, t^2)}{\partial t^1} + \Gamma^k_{ij}(t^1, t^2)\frac{\partial x^i(t^1, t^2)}{\partial t^1} X^j(t^1, t^2),$$

where $X^k(t^1, t^2)$ are the components of the vector $X(t^1, t^2)$; also

$$x^i(t^1, t^2) = x^i(\varphi(t^1, t^2)),$$

and

$$\Gamma^i_{jk}(t^1, t^2) = \Gamma^i_{jk}(\varphi(t^1, t^2)).$$

In particular, we obtain from this the following expressions for the components $\left(\frac{\nabla}{\partial t^1}\frac{\partial\varphi}{\partial t^2}\right)^k(t^1, t^2)$ of the vector $\frac{\nabla}{\partial t^1}\frac{\partial\varphi}{\partial t^2}(t^1, t^2)$:

$$\left(\frac{\nabla}{\partial t^1}\frac{\partial\varphi}{\partial t^2}\right)^k(t^1, t^2) = \\ = \frac{\partial^2 x^k(t^1, t^2)}{\partial t^1\,\partial t^2} + \Gamma^k_{ij}(t^1, t^2)\frac{\partial x^i(t^1, t^2)}{\partial t^1}\frac{\partial x^j(t^1, t^2)}{\partial t^2}.$$

(Here, we recall that the components of the vector $\frac{\partial\varphi}{\partial t^2}(t^1, t^2)$ are the derivatives $\frac{\partial x^i(t^1, t^2)}{\partial t^2}$.) Analogously,

$$\left(\frac{\nabla}{\partial t^2}\frac{\partial\varphi}{\partial t^1}\right)^k(t^1, t^2) = \\ = \frac{\partial^2 x^k(t^1, t^2)}{\partial t^2\,\partial t^1} + \Gamma^k_{ij}(t^1, t^2)\frac{\partial x^i(t^1, t^2)}{\partial t^2}\frac{\partial x^j(t^1, t^2)}{\partial t^1}.$$

If $K = 0$, we have $\Gamma^k_{ij} = \Gamma^k_{ji}$ and these vectors coincide. This proves that

If $K = 0$, *the formula*

$$\frac{\nabla}{\partial t^1}\frac{\partial\varphi}{\partial t^2} = \frac{\nabla}{\partial t^2}\frac{\partial\varphi}{\partial t^1} \tag{2}$$

is valid for an arbitrary surface $\varphi(t^1, t^2)$.

Returning to an arbitrary vector field $X(t^1, t^2)$ on the surface φ, let us consider the field $\frac{\nabla}{\partial t^2}\frac{\nabla X}{\partial t^1}(t^1, t^2)$. Its components are of the form

$$\left(\frac{\nabla}{\partial t^2}\frac{\nabla X}{\partial t^1}\right)^k (t^1, t^2) =$$
$$= \frac{\partial^2 X^k(t^1, t^2)}{\partial t^2\, \partial t^1} + \frac{\partial \Gamma^k_{ij}}{\partial x^l}(t^1, t^2)\frac{\partial x^l(t^1, t^2)}{\partial t^2}\frac{\partial x^i(t^1, t^2)}{\partial t^1} X^j(t^1, t^2) +$$
$$+ \Gamma^k_{ij}(t^1, t^2)\frac{\partial^2 x^i(t^1, t^2)}{\partial t^2\, \partial t^1} X^j(t^1, t^2) +$$
$$+ \Gamma^k_{ij}(t^1, t^2)\frac{\partial x^i(t^1, t^2)}{\partial t^1}\frac{\partial X^j(t^1, t^2)}{\partial t^2} +$$
$$+ \Gamma^k_{ls}(t^1, t^2)\frac{\partial x^l(t^1, t^2)}{\partial t^2}\left(\frac{\partial X^s(t^1, t^2)}{\partial t^1} + \Gamma^s_{ij}(t^1, t^2)\frac{\partial x^i(t_1, t_2)}{\partial t^1} X^j(t^1, t^2)\right),$$

where

$$\frac{\partial \Gamma^k_{ij}}{\partial x^l}(t^1, t^2) = \frac{\partial \Gamma^k_{ij}}{\partial x^l}(\varphi(t^1, t^2)).$$

If we subtract from this formula the analogous formula for

$$\left(\frac{\nabla}{\partial t^1}\frac{\nabla X}{\partial t^2}\right)^k (t^1, t^2),$$

we obtain

$$\left(\frac{\nabla}{\partial t^2}\frac{\nabla X}{\partial t^1}\right)^k (t^1, t^2) - \left(\frac{\nabla}{\partial t^1}\frac{\nabla X}{\partial t^2}\right)^k (t^1, t^2) =$$
$$= \Big(\Gamma^k_{is}(t^1, t^2)\Gamma^s_{jl}(t^1, t^2) - \Gamma^k_{js}(t^1, t^2)\Gamma^s_{il}(t^1, t^2) +$$
$$+ \frac{\partial \Gamma^k_{jl}}{\partial x^i}(t^1, t^2) - \frac{\partial \Gamma^k_{il}}{\partial x^j}(t^1, t^2)\Big)\frac{\partial x^i(t^1, t^2)}{\partial t^2}\frac{\partial x^j(t^1, t^2)}{\partial t^1} X^l(t^1, t^2),$$

that is (cf. formula (2′) of section 2),

$$\left(\frac{\nabla}{\partial t^2}\frac{\nabla X}{\partial t^1}\right)^k (t^1, t^2) - \left(\frac{\nabla}{\partial t^1}\frac{\partial X}{\partial t^2}\right)^k (t^1, t^2) =$$
$$= R^k_{lij}(t^1, t^2)\frac{\partial x^i(t^1, t^2)}{\partial t^2}\frac{\partial x^j(t^1, t^2)}{\partial t^1} X^l(t^1, t^2),$$

where

$$R^k_{lij}(t^1, t^2) = R^k_{lij}(\varphi(t^1, t^2)).$$

Turning from the components to the vector fields, we obtain this result

$$\frac{\nabla}{\partial t^2}\frac{\nabla X}{\partial t^1}(t^1, t^2) - \frac{\nabla}{\partial t^1}\frac{\nabla X}{\partial t^2}(t^1, t^2) = R_{\varphi(t^1, t^2)}\left(\frac{\partial \varphi}{\partial t^2}, \frac{\partial \varphi}{\partial t^1}\right) X(t^1, t^2),$$

that is,

$$\frac{\nabla}{\partial t^2}\frac{\nabla}{\partial t^1} - \frac{\nabla}{\partial t^1}\frac{\nabla}{\partial t^2} = R_{\varphi(t^1, t^2)}\left(\frac{\partial \varphi}{\partial t^2}, \frac{\partial \varphi}{\partial t^1}\right), \tag{3}$$

where both sides of the equation are treated as operators acting on the vector fields $X(t^1, t^2)$.

Suppose, finally, that

$$\gamma(t) = \varphi(t^1(t), t^2(t))$$

is an arbitrary curve belonging to the surface φ. On this curve, the vector field $X(t^1, t^2)$ defines the field

$$X(t) = X(t^1(t), t^2(t)).$$

An easy calculation (which we omit) shows that the following formula (analogous to the familiar formula in elementary analysis) is valid for the covariant derivative $\frac{\nabla X}{dt}(t)$ of this field:

$$\frac{\nabla X}{dt}(t) = \frac{\partial X}{\partial t^1}(t^1(t), t^2(t))\,\dot{t}^1(t) + \frac{\partial X}{\partial t^2}(t^1(t), t^2(t))\,\dot{t}^2(t).$$

Obviously, the corresponding formula

$$\frac{\nabla X}{\partial t}(t) = \frac{\partial X}{\partial t^i}(t^1(t), \ldots, t^n(t))\,\dot{t}^i(t) \tag{4}$$

is valid for arbitrary n.

4. Parallel Translation Along a Curve

We shall say that a vector field $X(t)$ on a curve $\gamma(t)$, where $t \in I$, is a field of parallel vectors if

$$\frac{\nabla X}{dt}(t) = 0 \quad \text{for every } t \in I.$$

According to formula (1) of section 3,

A field $X(t)$ is a field of parallel vectors if and only if, for an arbitrary system of local coordinates $x^1, \ldots, x^m$, its components $X^k(t) = X(t)\,x^k$ satisfy the equations

$$\frac{dX^k(t)}{dt} + \Gamma^k_{ij}(t)\,\dot{x}^i(t)\,X^j(t) = 0, \qquad k = 1, \ldots, m. \tag{1}$$

Suppose now that $p = \gamma(t_0)$ and $q = \gamma(t_1)$ are two arbitrary points on the curve $\gamma(t)$, where $t \in I$. We shall say that a vector $B \in M_q$ is obtained from the vector $A \in M_p$ by a parallel displacement along the curve γ if, for some interval $J \subset I$ containing the points t_0 and t_1, there exists a field $X(t)$ of parallel vectors on the curve $\gamma|_J$ such that

$$X(t_0) = A, \quad X(t_1) = B.$$

In this case, we shall write

$$B = \underset{q}{\tau^p} A.$$

It follows immediately from this definition that the parallel displacement $\tau = \underset{q}{\tau^p}$ is a transitive operation:

For any three points $p = \gamma(t_0)$, $q = \gamma(t_1)$, ***and*** $r = \gamma(t_2)$ ***on the curve*** γ ***and for any three vectors*** $A \in M_p$, $B \in M_q$, ***and*** $C \in M_r$, ***the relations*** $B = \underset{q}{\tau^p} A$ ***and*** $C = \underset{r}{\tau^q} B$ ***imply the relation*** $C = \underset{r}{\tau^p} A$.

Let us now show that

For an arbitrary vector $A \in M_p$, ***the vector*** $\underset{q}{\tau^p} A \in M_q$ always exists and is uniquely defined; also the mapping

$$\underset{q}{\tau^p} : M_p \to M_q$$

is an isomorphism between linear spaces.

By virtue of the transitivity of the displacement τ, it will be sufficient to prove this assertion for the case in which the curve γ lies entirely within a coordinate neighborhood. Also, since the mapping $\underset{p}{\tau^p}$ is the identity mapping, the fact that the mapping $\underset{q}{\tau^p}$ is isomorphic follows from its linearity and transitivity (since $\underset{p}{\tau^q} \circ \underset{q}{\tau^p} = \underset{p}{\tau^p}$).

Keeping this in mind, let us look at equations (1), which are satisfied by the components $X^k(t)$ of an arbitrary field of parallel vectors on the curve γ. According to a familiar theorem in the theory of differential equations, there exist smooth functions $g^i(t;\, a^1, \ldots, a^m)$ defined on the interval I, depending linearly on the m real parameters $a^1, \ldots, a^m$ and such that

(1) for arbitrary $a^1, \ldots, a^m$, the functions $X^i(t) = g^i(t;\, a^1, \ldots, a^m)$ satisfy equations (1);

(2) $g^i(t_0;\, a^1, \ldots, a^m) = a^i$;

Furthermore, the functions g^i are uniquely defined by these conditions. In geometrical language, this theorem states that, for arbitrary $a^1, \ldots, a^m$, the formula

$$X(t;\, a^1, \ldots, a^m) = g^i(t;\, a^1, \ldots, a^m)\left(\frac{\partial}{\partial x^i}\right)_{\gamma(t)}$$

defines on γ a field $X(t;\, a^1, \ldots, a^m)$ of parallel vectors such that

$$X(t_0;\, a^1, \ldots, a^m) = a^i\left(\frac{\partial}{\partial x^i}\right)_{\gamma(t_0)}.$$

Since the vector $a^i\left(\frac{\partial}{\partial x^i}\right)_{\gamma(t_0)}$ ranges over the entire space $M_{\gamma(t_0)} = M_p$, it follows that the mapping τ is defined on the entire space M_p. Since the functions g^i are unambiguously defined by conditions (1) and (2), the mapping τ is single-valued. Since the functions g^i depend linearly on $a^1, \ldots, a^m$, the mapping τ is linear.

This completes the proof of the assertion made above.

The operation $\tau = \underset{q}{\tau^p}$ parallel displacement along a curve γ from the point p to the point q can be defined not only for vectors but also for tensors of arbitrary type $(r, s) \neq (0, 0)$.

Suppose first that α is any tensor of type (0, 1) at the point p, that is, a covector $\alpha : M_p \to R$. Let us define the covector $\tau\alpha : M_q \to R$ by setting, for an arbitrary vector $A \in M_q$,

$$\tau\alpha(A) = \alpha(\tau^{-1}A),$$

where $\tau^{-1} = \tau_p^q : M_q \to M_p$ is the parallel displacement along the curve γ from the point q to the point p inverse to the displacement $\tau = \tau_q^p$. Furthermore, for an arbitrary tensor $L \in \mathfrak{G}_s^r(p)$, where $r + s > 0$, we obtain

$$\tau L(\alpha^1, \ldots, \alpha^r; A_1, \ldots, A_s) = \\ = L(\tau^{-1}\alpha^1, \ldots, \tau^{-1}\alpha^r; \tau^{-1}A_1, \ldots, \tau^{-1}A_s),$$

where

$$\alpha^1, \ldots, \alpha^r \in \mathfrak{G}_1(q), \; A_1, \ldots, A_s \in \mathfrak{G}^1(q).$$

Obviously, the mapping

$$\tau : \mathfrak{G}_s^r(p) \to \mathfrak{G}_s^r(q), \qquad r + s > 0,$$

thus defined is also an isomorphism between linear spaces.

Now, let $T(t)$ denote a tensor field [of type $(r, s) \neq (0, 0)$] on the curve $\gamma(t)$, where $t \in I$. We shall say that this field consists of parallel tensors if, for arbitrary points $t_0, t_1 \in I$,

$$T(t_1) = \tau T(t_0),$$

where τ is the parallel translation along the curve γ from the point $\gamma(t_0)$ to the point $\gamma(t_1)$. Obviously, for vector fields, this definition is equivalent to the original one.

5. Covariant Differentiation of Tensor Fields

Let T denote an arbitrary tensor field on a space of affine connection M. Suppose that T is of type $(r, s) \neq (0, 0)$. Let X denote a vector field on M. Let us consider a point $p \in M$ at which $X_p \neq 0$. As we know (cf. section 12 of Chapter I), there exists an integral curve $\gamma(t)$ [exactly one] of the field X such that $\gamma(0) = p$. We define a tensor $(\nabla_X T)_p$ at the point p by setting

$$(\nabla_X T)_p = \lim_{t \to 0} \frac{\tau_t^{-1} T_{\gamma(t)} - T_p}{t}, \tag{1}$$

where τ_t is the parallel displacement along the curve γ from the point p to the point $\gamma(t)$.

Here, the limit is taken in the sense that, for an arbitrary system of r covectors and s vectors at the point p, the value of the tensor $(\nabla_X T)_p$ on this system is equal to the limit of values of the tensor $\frac{1}{t}(\tau_t^{-1} T_{\gamma(t)} - T_p)$. In particular, this means that, for an arbitrary basis for the space M_p, the components of the tensor $(\nabla_X T)_p$ are the limits of the corresponding components of the tensor $\frac{1}{t}(\tau_t^{-1}T_{\gamma(t)} - T_p)$.

If $X_p = 0$, let us define

$$(\nabla_X T)_p = 0.$$

Below, we shall show that:

(1) The mapping** $p \to (\nabla_X T)_p$ **satisfies the smoothness condition; that is,** $\nabla_X T$ **is a tensor field (of type** (r, s)**) on** M**.

We shall say that this field $\nabla_X T$ is obtained from the field T by covariant differentiation along the vector field X. This terminology is justified by the fact that

(2) For the vector field** T**, the operation** ∇_X **coincides with the operation** ${}^X\Delta$ **corresponding in the affine connection** ∇ **to the field** X**.

Now, let $\gamma(t)$, for $t \in I$, denote an arbitrary regular curve in a space M of affine connection and let $T(t)$ denote an arbitrary tensor field (of type (r, s)) on that curve. As we know, the field $T(t)$ on an arbitrary elementary segment $\gamma|_C$ of the curve γ can be extended to some field T on M. Let us define on $\gamma|_C$ a tensor field $\frac{\nabla T}{dt}(t)$ by

$$\frac{\nabla T}{dt}(t) = (\nabla_A T)_{\gamma(t)} \quad \text{for every} \quad t \in C,$$

where A is the extension to M of the field $\dot{\gamma}(t)$ (cf. the analogous definition for vector fields in section 3). As we shall show,

(3) The field** $\frac{\nabla T}{dt}(t)$ **is independent of the choice of extensions** T **and** A**; that is, it is defined exclusively by the field** $T(t)$ **and the curve** $\gamma(t)$**.

Since an arbitrary point on the curve $\gamma(t)$ belongs to some elementary segment, this construction for arbitrary $t \in I$ unambiguously defines the tensor $\frac{\nabla T}{dt}(t)$. Furthermore, as one can easily see, the mapping $t \to \frac{\nabla T}{dt}(t)$ satisfies the smoothness condition; that is, the tensors $\frac{\nabla T}{dt}(t)$ constitute a tensor field on the curve $\gamma(t)$. We shall say that this field is obtained from the field $T(t)$ by covariant differentiation along the curve $\gamma(t)$.

It turns out that

(4) A field** $T(t)$ **is a field of parallel tensors if and only if

$$\frac{\nabla T}{dt}(t) = 0 \quad \text{for all} \quad t \in I.$$

Thus, the parallelism of arbitrary tensors can be characterized in the same way as the parallelism of vectors (cf. section 4).

To prove assertions (1)-(4), let us first find conditions that the components of tensors $T(t)$ that are parallel on a curve $\gamma(t)$ must satisfy (under the assumption that the curve $\gamma: I \to M$ lies entirely in some coordinate neighborhood U).

To begin with, suppose that $r = 0$ and $s = 1$; that is, suppose that the tensors $T(t)$ are covectors $\alpha(t)$. Then, in an arbitrary system of coordinates $x^1, \ldots, x^m$ defined in the neighborhood U,

$$\alpha(t) = \alpha_k(t)\,(dx^k)_{\gamma(t)},$$

where the $\alpha_k(t)$ are functions that are smooth on I. By definition, the condition $\alpha(t_1) = \tau\alpha(t_0)$ stating that the covectors $\alpha(t)$ are parallel at points $t_0, t_1 \in I$ is equivalent to satisfaction of the equation

$$\alpha(t_1)(X(t_1)) = \alpha(t_0)(X(t_0)).$$

for an arbitrary field

$$X(t) = X^k(t)\left(\frac{\partial}{\partial x^k}\right)_{\gamma(t)}$$

of vectors parallel on the curve $\gamma(t)$. Thus, the covectors $\alpha(t)$ are parallel on $\gamma(t)$ if and only if, for an arbitrary field $X(t)$ of vectors parallel on γ, the number

$$\alpha(t)(X(t)) = \alpha_k(t)\,X^k(t)$$

is independent of t. If we differentiate the equation

$$\alpha_k(t)\,X^k(t) = \text{const}$$

with respect to t and keep in mind formulas (1) of section 4, we see that this equation is equivalent to the relation

$$\left(\dot{\alpha}_k(t) - \Gamma^j_{ik}(t)\,\dot{x}^i(t)\,\alpha_j(t)\right) X^k(t) = 0.$$

Since the numbers $X^k(t)$ can assume arbitrary values for each $t \in I$, this equation means that

$$\frac{d\alpha_k(t)}{dt} - \Gamma^j_{ik}(t)\,\dot{x}^i(t)\,\alpha_j(t) = 0, \qquad k = 1, \ldots, m. \tag{2}$$

Thus, we have proven that

A field $\alpha(t)$ is a field of parallel vectors if and only if its components $\alpha_k(t)$ satisfy equations (2).

Suppose now that $r = 1$ and $s = 1$, so that

$$T(t) = T^i_j(t)\,(dx^j)_{\gamma(t)} \otimes \left(\frac{\partial}{\partial x^i}\right)_{\gamma(t)}.$$

Just for the case $r = 0$ and $s = 1$, we easily see that the tensors $T(t)$ are parallel on $\gamma(t)$ if and only if, for arbitrary covectors $\alpha(t)$

and vectors $X(t)$ that are parallel on $\gamma(t)$, the function

$$T(t)(\alpha(t);\ X(t)) = T^i_j(t)\,\alpha_i(t)\,X^j(t)$$

is independent of t. If we differentiate this function with respect to t and remember formulas (1) of section 4 and (2) of the present section, we easily see that

A field $T(t)$ of type (1, 1) is a field of parallel tensors if and only if its components $T^i_j(t)$ in an arbitrary system of local coordinates satisfies the equations

$$\frac{dT^i_j(t)}{dt} + \Gamma^i_{kl}(t)\,\dot{x}^k(t)\,T^l_j(t) - \Gamma^l_{kj}(t)\,\dot{x}^k(t)\,T^i_l(t) = 0. \tag{3}$$

The case of arbitrary r and s is developed in a completely analogous fashion. Here, the equations defining the components of the field $T(t)$ contain r terms of "vector" type [analogous to the second term in formula (3)] and s terms of "covector" type [analogous to the third term in formula (3)].

Let us now calculate the components of the tensor $(\nabla_X T)_p$ defined by formula (1). To simplify our calculations, we shall assume that $r = 1$ and $s = 1$ since consideration of the general case would complicate the presentation unnecessarily.

Let us fix some value $t = t_0$ of the parameter t and perform a parallel translation of the tensor $T_{\gamma(t_0)}$ along the curve γ into each point $\gamma(t)$, where $0 \leqslant t \leqslant t_0$. When we do this, we define a field of parallel tensors $S(t)$ on the segment $\gamma|_{[0,\,t_0]}$. Here, the equations

$$S(t_0) = T_{\gamma(t_0)},\quad S(0) = \tau_{t_0}^{-1} T_{\gamma(t_0)}$$

are satisfied. Let $S^i_j(t)$ and $T^i_j(t)$ denote, respectively, the components of the tensor $S(t)$ and $T(t)$ relative to the basis

$$\left(\frac{\partial}{\partial x^1}\right)_{\gamma(t)},\ \ldots,\ \left(\frac{\partial}{\partial x^m}\right)_{\gamma(t)}.$$

Here, $x^1, \ldots, x^m$ are local coordinates at the point p. We assume t_0 is sufficiently small that the entire segment $\gamma|_{[0,\,t_0]}$ lies in a coordinate neighborhood of the system of local coordinates $x^1, \ldots, x^m$. Then, since $S^i_j(t_0) = T^i_j(t_0)$, the tensor

$$\frac{\tau_{t_0}^{-1} T_{\gamma(t_0)} - T_p}{t_0} = \frac{S(0) - T_{\gamma(0)}}{t_0}$$

will have the components

$$\frac{S^i_j(0) - T^i_j(0)}{t_0} = \frac{S^i_j(0) - S^i_j(t_0)}{t_0} + \frac{T^i_j(t_0) - T^i_j(0)}{t_0}.$$

According to Lagrange's theorem, the first term on the right is equal to $-\dot{S}^i_j(t^*)$, where $0 \leqslant t^* \leqslant t_0$ and hence approaches $-\dot{S}^i_j(0)$ as

$t_0 \to 0$. The second term approaches $\dot{T}^i_j(0)$ as $t_0 \to 0$. Thus, we obtain the following expression for the components $((\nabla_X T)_p)^i_j$ of the tensor $(\nabla_X T)_p$ (relative to the basis $\left(\frac{\partial}{\partial x^1}\right)_p, \ldots, \left(\frac{\partial}{\partial x^m}\right)_p$):

$$((\nabla_X T)_p)^i_j = -\dot{S}^i_j(0) + \dot{T}^i_j(0).$$

On the other hand, the components $S^i_j(t)$ satisfy equations (3), that is, the equations

$$\dot{S}^i_j(t) + \Gamma^i_{kl}(t)\,\dot{x}^k(t)\,S^l_j(t) - \Gamma^l_{kj}(t)\,\dot{x}^k(t)\,S^i_l(t) = 0.$$

From this it follows that

$$-\dot{S}^i_j(t_0) = \Gamma^i_{kl}(t_0)\,\dot{x}^k(t_0)\,T^l_j(t_0) - \Gamma^l_{kj}(t_0)\,\dot{x}^k(t_0)\,T^i_l(t_0),$$

since $S(t_0) = T(t_0)$. Introducing the components $X^k = Xx^k$ and $T^l_j = T\left(dx^l, \frac{\partial}{\partial x^j}\right)$ of the fields X and T and taking the limit as $t_0 \to 0$, we obtain

$$\begin{aligned} -\dot{S}^i_j(0) &= \Gamma^i_{kl}(0)\,\dot{x}^k(0)\,T^l_j(0) - \Gamma^l_{kj}(0)\,\dot{x}^k(0)\,T^i_l(0) = \\ &= (\Gamma^i_{kl}X^kT^l_j - \Gamma^l_{kj}X^kT^i_l)_p, \end{aligned}$$

where the index p means that all the quantities in the parentheses are evaluated at the point p. (We recall that $\gamma(t)$ is an integral curve of the field X and hence $\dot{x}^k(0) = X^k_p$).)

Furthermore, since

$$T^i_j(t) = T^i_j(\gamma(t)),$$

we obtain from the chain rule

$$\frac{dT^i_j(t)}{dt} = \left(\frac{\partial T^i_j}{\partial x^k}\right)_{\gamma(t)} \frac{dx^k(t)}{dt},$$

the relation

$$\dot{T}^i_j(0) = \left(\frac{\partial T^i_j}{\partial x^k} X^k\right)_p.$$

Thus, we obtain the following final expression for the components of the tensor $(\nabla_X T)_p$ (with the index p omitted):

$$(\nabla_X T)^i_j = \left(\frac{\partial T^i_j}{\partial x^k} + \Gamma^i_{kl}T^l_j - \Gamma^l_{kj}T^i_l\right)X^k. \tag{4}$$

Up to now, we have assumed that $X_p \neq 0$. However, it is clear that formula (4) remains valid for $X_p = 0$ [because in this case, by hypothesis, $(\nabla_X T)_p = 0$)]. Assertions (1)-(4) now follow directly from formula (4).

Specifically, to prove assertion (1), we need only note that all the functions on the right-hand side of formula (4) are smooth at the point p.

The analog of formula (4) for vector fields is of the form

$$(\nabla_X Y)^i = \left(\frac{\partial Y^i}{\partial x^k} + \Gamma^i_{kl} Y^l\right) X^k.$$

This formula coincides with formula (1) of section 1. This completes the proof of assertion (2).

Furthermore, it follows immediately from formula (4) [cf. the corresponding reasoning for vector fields in section 3] that the components $\frac{\nabla T^i_j}{dt}(t)$ of the field $\frac{\nabla T}{dt}(t)$ on an arbitrary elementary segment $\gamma|_C$ of the curve γ are given by the formula

$$\frac{\nabla T^i_j}{dt}(t) = \frac{dT^i_j(t)}{dt} + \Gamma^i_{kl}(t)\dot{x}^k(t) T^l_j(t) - \Gamma^l_{kj}(t)\dot{x}^k(t) T^i_l(t). \tag{5}$$

Since the right-hand side of formula (5) is independent of the choice of fields X and A, assertion (3) is proven.

Finally, if we compare formula (5) with formula (3), we immediately see that assertion (4) is valid.

Thus, assertions (1)-(4) are completely proven.

Up to now, we have assumed that $r+s>0$. Now, however, we shall consider the operation ∇_X on tensor fields of type (0, 0) as well, that is, on smooth functions $f \in \mathfrak{G}(M)$. We define the operation by

$$\nabla_X f = Xf, \quad X \in \mathfrak{G}^1(M), \quad f \in \mathfrak{G}(M).$$

Thus, the operation ∇_X is defined on all elements of the algebra $\mathfrak{O}(M)$. By definition, this operation conserves tensor type. Furthermore, as one can easily see,

The operation ∇_X is differentiation of the algebra $\mathfrak{O}(M)$ and it commutes with the contraction operations C^i_j.

Remark: The operation $\nabla_X = X$, as applied to functions, can be defined by a formula analogous to formula (1) for tensors. Specifically,

For an arbitrary function $f \in \mathfrak{G}(M)$ and an arbitrary field $X \in \mathfrak{G}^1(M)$, the number $X_p f$ is defined at every point $p \in M$ at which $X_p \neq 0$ by the formula

$$X_p f = \lim_{t \to 0} \frac{f(\gamma(t)) - f(p)}{t}, \tag{6}$$

where $\gamma(t)$ is the integral curve of the field X that passes through the point $p = \gamma(0)$.

Proof: The right-hand side of this formula is the ordinary derivative $df(t)/dt$ of the function $f(t) = f(\gamma(t))$ at the point $t=0$. But, from the chain rule (cf. section 12 of Chapter 1),

$$\frac{df(t)}{dt} = \left(\frac{\partial f}{\partial x^i}\right)_{\gamma(t)} \frac{dx^i(t)}{dt} = \dot{\gamma}(t) f.$$

If we now set $t = 0$ and remember that $\dot{\gamma}(0) = X_p$, we see immediately the validity of the assertion to be proven.

In what follows, we shall need the following somewhat more general assertion:

For an arbitrary function $f \in \mathfrak{G}(p)$ *and an arbitrary vector* $C \in M_p$,

$$Cf = \lim_{t \to 0} \frac{f(\gamma(t)) - f(p)}{t},$$

where $\gamma(t)$ *is an arbitrary smooth (not necessarily regular) curve such that*

$$\gamma(0) = p, \qquad \dot{\gamma}(0) = C.$$

To prove this assertion, we need only repeat verbatim the preceding reasoning.

In particular, formula (6) is suitable for determining the numbers $X_p f$ and it is suitable in the case $X_p = 0$ if we understand for the curve $\gamma(t)$ the degenerate curve $R \to p$. The same holds obviously for formula (1) defining the tensor $(\nabla_X T)_p$.

6. Geodesics

A smooth curve $\gamma: I \to M$ in a space of affine connection M is called a *geodesic* if the vectors of the field $\dot{\gamma}(t)$ are parallel on γ, that is, if

$$\frac{\nabla \dot{\gamma}}{dt}(t) = 0 \qquad \text{for all} \quad t \in I.$$

This means that the functions $x^k(t)$ defining the geodesic $\gamma(t)$ parametrically in some system of local coordinates satisfy the equations

$$\frac{d^2 x^k(t)}{dt^2} + \Gamma_{ij}^k(t) \frac{dx^i(t)}{dt} \frac{dx^j(t)}{dt} = 0. \tag{1}$$

Obviously,

For an arbitrary geodesic $\gamma(t)$ *and arbitrary numbers* a *and* b, *the curve* $\gamma(at + b)$ *is also a geodesic.*

We shall say that the geodesic $\gamma(at + b)$ is obtained from the geodesic $\gamma(t)$ by a linear transformtion of parameter.

In what follows, we shall need the following simple remark:

Suppose that a continuous curve $\gamma(t)$ *is twice continuously differentiable, that is, for every point* p *on it and for an arbitrary system of local coordinates* $x^1, \ldots, x^m$ *at the point* p, *the functions* $x^1(t), \ldots, x^m(t)$, *which define the curve* $\gamma(t)$ *parametrically in the corresponding coordinate neighborhood, have continuous second derivatives. Then, if these functions satisfy equations (1), the curve* $\gamma(t)$ *is smooth and therefore is a geodesic.*

Proof: We can determine successively all the derivatives of the functions $x^i(t)$ by differentiating equation (1) the necessary number of times.

It is also useful to keep in mind the fact that the property of "being a geodesic" is a local property; that is,

A curve $\gamma(t)$ *in a space* M *is a geodesic if it is locally a geodesic, that is, if every point of it lies in a neighborhood such that the segment of the curve* $\gamma(t)$ *contained in that neighborhood is a geodesic.*

We note also the following:

Any geodesic is either a regular or a degenerate curve.

To see this, note that if a vector $\dot{\gamma}(t)$ is equal to 0 or at least one value of the parameter t, it is also 0 for all values of t (because parallel translation is an isomorphism).

It follows immediately from the theorem on the existence and uniqueness of solutions of systems of differential equations [as applied to the system (1)] that

For an arbitrary point $p \in M$, *an arbitrary vector* $A \in M_p$, *and an arbitrary number* t_0, *there exists a geodesic* $\gamma(t)$ *defined in some neighborhood of the point* t_0 *such that*

$$\gamma(t_0) = p, \qquad \dot{\gamma}(t_0) = A.$$

Here, any two geodesics coincide in some neighborhood of the point t_0.

Furthermore, as one can easily see,

If two geodesics $\gamma_1(t)$ *and* $\gamma_2(t)$ *defined, respectively, on (intersecting) intervals* I_1 *and* I_2 *have the property that for some point* $t_0 \in I_1 \cap I_2$,

$$\gamma_1(t_0) = \gamma_2(t_0), \qquad \dot{\gamma}_1(t_0) = \dot{\gamma}_2(t_0),$$

then the geodesics $\gamma_1(t)$ *and* $\gamma_2(t)$ *coincide on the entire interval* $I = I_1 \cap I_2$.

Proof: Let C be the set of points $t \in I$ at which

$$\gamma_1(t) = \gamma_2(t), \qquad \dot{\gamma}_1(t) = \dot{\gamma}_2(t).$$

According to the preceding assertion, this set is open. On the other hand, since the curves $\gamma_1(t)$ and $\gamma_2(t)$ and the fields $\dot{\gamma}_1(t)$ and $\dot{\gamma}_2(t)$ are continuous and since the space M is a Hausdorff space, this set is closed (in I). Since it is nonempty (because $t_0 \in C$) and the interval I is connected, this is possible only if $C = I$.

A geodesic γ is called ***maximal*** if it is not the restriction of any other geodesic defined on a larger interval of the real axis. From the assertion just proven, it follows immediately that

For an arbitrary point $p \in M$, *an arbitrary vector* $A \in M_p$, *and an arbitrary number* t_0, *there exists exactly one maximal geodesic* γ *defined on some interval containing the point* t_0 *and such that*

$$\gamma(t_0) = p, \qquad \dot{\gamma}(t_0) = A. \tag{2}$$

Proof: Let $\mathcal{G}$ denote the set of all geodesics $\gamma(t)$ that satisfy equations (2). Each of these geodesics is defined on some interval

I_γ that contains the point t_0. From what we have proved, the set $\mathcal{G}$ is nonempty and any two geodesics $\gamma_1, \gamma_2 \in \mathcal{G}$ coincide on the interval $I_{\gamma_1} \cap I_{\gamma_2}$. Therefore, the formula

$$\bar{\gamma}(t) = \gamma(t), \qquad \text{for} \quad t \in I_\gamma,$$

unambiguously defines some curve $\bar{\gamma}: I \to M$ on the interval

$$I = \bigcup_{\gamma \in \mathcal{G}} I_\gamma.$$

Obviously, this curve is the desired maximal geodesic. It is regular if $A \neq 0$ and it is a constant mapping of the real axis into the point p if $A = 0$.

In what follows, we shall take $t_0 = 0$. We shall denote by $\gamma_{p, A}$, or simply by γ_A when this will not cause misunderstanding, the maximal geodesic γ for which

$$\gamma(0) = p, \qquad \dot{\gamma}(0) = A.$$

It is easy to see that

For an arbitrary number a,

$$\gamma_{aA}(t) = \gamma_A(at). \tag{3}$$

To prove this, we need only note that $\dot{\gamma}_A(at)|_{t=0} = aA$ and use the uniqueness of the geodesics γ_A.

7. Normal Neighborhoods

Let D_p denote the set of those vectors A in the space M_p for which the geodesic γ_A is defined on an interval containing the point $t = 1$. For an arbitrary vector $A \in D_p$, we define

$$\operatorname{Exp} A = \gamma_A(1). \tag{1}$$

We shall call this mapping Exp: $D_p \to M$ the *exponential mapping*.

We shall say that a neighborhood V_0 of the zero vector of the space M_p is *normal* if

(1) it is "starlike", that is, if, for an arbitrary vector $A \in V_0$ and an arbitrary nonnegative number $a \leqslant 1$, the vector aA belongs to the neighborhood V_0;

(2) the mapping Exp is defined on V_0, that is, $V_0 \subset D_p$;

(3) the mapping Exp on V_0 is a diffeomorphism of that neighborhood onto some neighborhood $V(p)$ of the point p.

We shall refer to the neighborhood

$$V(p) = \operatorname{Exp} V_0$$

mentioned in condition (3) as a *normal neighborhood* of the point $p \in M$.

It turns out that

Every point p_0 *in a space* M *of affine connection has a normal neighborhood* $V(p_0)$.

Proof: Let us choose a (fixed) system of local coordinates $x^1, \ldots, x^m$ at the point p_0 such that

$$x^1(p_0) = \ldots = x^m(p_0) = 0.$$

By decreasing (if necessary) the corresponding coordinate neighborhood U, we can assume that the coordinate homeomorphism $\xi: U \to R^m$ maps the neighborhood U onto the open ball

$$(t^1)^2 + \ldots + (t^m)^2 < c_1^2$$

of radius c_1 with center at the point $0 \in R^m$.

Now, let us introduce a metric in the neighborhood U by taking for the distance $\rho(p, q)$ between any two points p and q the Euclidean distance

$$\sqrt{\sum_{i=1}^{m} (x^i(p) - x^i(q))^2}$$

between corresponding points $\xi(p)$ and $\xi(q)$ in the space R^m. For any point $p \in U$, we denote by $U_\delta(p)$ the δ-neighborhood of p in this metric, that is, the set of all points $q \in U$ at which $\rho(p, q) < \delta$. In particular,

$$U_{c_1}(p_0) = U.$$

Furthermore, in each of the spaces M_p, where $p \in U$, we introduce a scalar product by assuming the basis

$$\left(\frac{\partial}{\partial x^1}\right)_p, \ldots, \left(\frac{\partial}{\partial x^m}\right)_p$$

of this space to be orthonormal. This enables us to speak of the lengths of vectors in the space M_p.

Let us apply to equations (1) of section 6 the theorem on the dependence of the solutions of a differential equation on the initial conditions. According to this theorem, for arbitrary $c_1 > 0$ there exist a $c > 0$ and smooth functions

$$x^i(t;\ \xi^1, \ldots, \xi^m;\ a^1, \ldots, a^m)$$

of the $2m + 1$ variables $t, \xi^1, \ldots, \xi^m, a^1, \ldots, a^m$ such that, for arbitrary fixed $\xi^1, \ldots, \xi^m, a^1, \ldots, a^m$ satisfying the inequalities

$$(\xi^1)^2 + \ldots + (\xi^m)^2 < c^2, \quad (a^1)^2 + \ldots + (a^m)^2 < c^2,$$

the functions

$$x^i(t) = x^i(t;\ \xi^1. \ldots, \xi^m;\ a^1, \ldots, a^m)$$

satisfy for $|t| \leqslant 2c$ equations (1) of section 6 with initial conditions

$$x^i(0) = \xi^i, \quad \dot{x}^i(0) = a^i, \quad i = 1, \ldots, m$$

and, in addition, have the property that

$$(x^1(t))^2 + \ldots + (x^m(t)^2) < c_1^2 \quad \text{for} \quad |t| \leqslant 2c.$$

In the language of curves in the space M, this means that, for an arbitrary point $p \in U_c(p_0)$ and an arbitrary vector $A \in M_p$ such that $|A| < c$, the maximal geodesic $\gamma_{p,A}$ for $|t| \leqslant 2c$ is contained in the neighborhood U and depends "smoothly" on the point p and the vector A in the sense that the functions

$$x^i(t;\ p;\ A) = x^i(t;\ x^1(p), \ldots, x^m(p);\ a^1, \ldots, a^m),$$

defining the geodesic $\gamma_{p,A}$ parametrically on the interval $|t| \leqslant 2c$ depend smoothly on the coordinates $x^1(p), \ldots x^m(p)$ and $a^1, \ldots, a^m$ of the point p and on the vector A.

In accordance with formula (3) of section 6, for an arbitrary number a at which the expression $x^i(at;\ p;\ A)$ is defined,

$$x^i(at;\ p;\ A) = x^i(t;\ p;\ aA). \tag{2}$$

In particular, this equation holds for $|a| \leqslant 1$.

If we differentiate equation (2) twice with respect to a and set

$$D^2 x^i(t;\ p;\ A) = \frac{\partial^2}{\partial t^2} x^i(t;\ p;\ A),$$

$$D_{jk} x^i(t;\ p;\ A) = \frac{\partial^2}{\partial a^j\, \partial a^k} x^i(t;\ p;\ A), \quad j,\ k = 1, \ldots, m,$$

we obtain

$$t^2 D^2 x^i(at;\ p;\ A) = a^j a^k D_{jk} x^i(t;\ p;\ aA).$$

From this and from Maclaurin's formula

$$f(t) = f(0) + \dot{f}(0)\, t + \frac{1}{2} \ddot{f}(\theta t)\, t^2, \quad 0 < \theta < 1,$$

as applied to the function $f(t) = x^i(t;\ p;\ A)$, it follows immediately that

$$x^i(t;\ p;\ A) = x^i(p) + a^i t + \frac{1}{2} a^j a^k D_{jk} x^i(t;\ p;\ \theta A),$$

since, in accordance with the initial conditions,

$$x^i(0;\ p;\ A) = x^i(p), \quad \frac{\partial x^i}{\partial t}(0;\ p;\ A) = a^i.$$

Consequently,

$$\left(\frac{\partial x^i(t;\ p;\ A)}{\partial a^j}\right)_{A=0} = \delta^i_j t$$

and, in particular,

$$\left(\frac{\partial x^i(c;\ p;\ A)}{\partial a^j}\right)_{A=0} = \delta^i_j c, \quad i = 1, \ldots, m. \tag{3}$$

By hypothesis, for any of these vectors A, the point $\gamma_A(c) = \gamma_{p,\,A}(c)$ is defined. Therefore, for the vector cA, the point $\gamma_{cA}(1) = \mathrm{Exp}\,(cA)$ is defined, so that $cA \in D_p$. This means that the mapping Exp is defined on some neighborhood V of the vector $0 \in M_p$ (specifically, on the neighborhood consisting of vectors A such that $|A| \leqslant c$).

For an arbitrary vector A in this neighborhood, the point Exp $A = \gamma_{\frac{A}{c}}(c)$ has coordinates

$$x^1\left(c;\ p;\ \frac{A}{c}\right), \ldots, x^m\left(c;\ p;\ \frac{A}{c}\right). \tag{4}$$

This means that the composite mapping $\xi \circ$ Exp of the neighborhood $V \subset M_p$ into the space R^m, where ξ is the coordinate homemorphism corresponding to the system of local coordinates $x^1, \ldots, x^m$, is defined by the functions (4). In accordance with formulas (3), the Jacobian of this mapping is equal to one at the point $A = 0$. Therefore (by virtue of the implicit-function theorem), there exists a positive number $\varepsilon_p \leqslant c^2$ such that the mapping $\xi \circ$ Exp is a diffeomorphism on the neighborhood V_0 of the vector $0 \in M_p$ consisting of those vectors $A \in M_p$ such that $|A| < \varepsilon_p$. Consequently, the mapping Exp is also a diffeomorphism on V_0. Since, obviously, the neighborhood V_0 is starlike, we have proved that

An arbitrary point $p \in U_c(p_0)$ has a normal neighborhood $V(p) =$ Exp (V_0).

In particular, the point p_0 has a normal neighborhood. This completes the proof of the assertion made on page 68.

Remark: By analyzing the familiar implicit-function theorem, one can easily show that

The "radii" ε_p of the normal neighborhoods V_0 constructed above can be chosen so that they depend continuously on the point p.

It follows immediately from this remark that

There exists a continuous function $\eta(p)$ on the neighborhood $U_c(p_0)$ such that, for an arbitrary point $p \in U_c(p_0)$, the spherical $\eta(p)$-neighborhood $U_{\eta(p)}(p)$ of that point p is contained in its normal neighborhood $V(p)$.

This is true because we may set

$$\eta(p) = \min_{|A| = \varepsilon_p} \sqrt{\sum_{i=1}^{m} \left(x^i\left(c;\ p;\ \frac{A}{c}\right) - x^i(p)\right)^2}.$$

Obviously, an arbitrary starlike neighborhood of the zero vector of the space M_p, if it is contained in a normal neighborhood, is itself a normal neighborhood. On the other hand, the starlike neighborhoods of the vector 0 in the tangent space M_p constitute a fundamental system of neighborhoods of that vector. From this it follows immediately that

The normal neighborhoods $V(p)$ constitute a fundamental system of neighborhoods of the point p in the manifold M.

The value of the normal neighborhoods is derived from the fact that

An arbitrary point q in a normal neighborhood $V(p)$ can be connected in $V(p)$ with the point p by a geodesic that is unique up to a linear transformation of parameter.

Specifically, the two points can be connected by γ_A, where A is a vector such that $\operatorname{Exp} A = q$.

One should not assume that every neighborhood (of a point p) with this property is a normal neighborhood. We can assert only the following:

Let V denote any neighborhood of a point p. Suppose V has the property that an arbitrary point $q \in V$ can be connected in V with the point p by a geodesic that is unique up to a linear transformation of parameter. Then, there exists a neighborhood V_0 of the vector $0 \in M_p$ such that

(1) the neighborhood V_0 is starlike;

(2) the mapping Exp is defined on the neighborhood V_0 and maps this neighborhood objectively onto the neighborhood V.

However, in general, the mapping Exp on the neighborhood V_0 is not a diffeomorphism (since the Jacobian of that mapping can vanish at some points of the neighborhood V_0).

To prove the above proposition, we need only note that the unique geodesic, say γ, that connects the point p with the point q must be a segment of some maximal geodesic $\gamma|_B$. Here, if $q = \gamma_B(c)$, then $q = \operatorname{Exp} A$, where $A = cB$. Thus, $\operatorname{Exp} D_p \supset V$. Let V_0 be the entire preimage of the neighborhood V under the mapping Exp. Because of the uniqueness of the geodesic γ, there exists only one vector $A \in V_0$ for which $\operatorname{Exp} A = q$. This shows that the neighborhood V_0 satisfies condition (2). It also satisfies condition (1) because, for arbitrary $a \in (0, 1)$ and an arbitrary vector $A \in V_0$, the point

$$\operatorname{Exp}(aA) = \gamma_{aA}(1) = \gamma_A(a)$$

belongs to the neighborhood V.

The mapping Exp on the neighborhood V_0 will obviously be a diffeomorphism if the neighborhood V is contained in some normal neighborhood $V(p)$ of the point p. Since we know that the geodesic connecting the point q with the point p is unique in this case, we have the following:

A neighborhood V contained in some normal neighborhood of a point p is a normal neighborhood of that point if an arbitrary point $q \in V$ can be connected in V to the point p by at least one geodesic.

Let us now suppose that a basis $A_1, \ldots, A_m$ is defined for the space M_p. Then, there is a linear isomorphism α of the Euclidean space R^m into the space M_p that maps the point $a = (a^1, \ldots, a^m)$ into the vector $a^i A_i$. Suppose that

$$\mathrm{E} = \mathrm{Exp} \circ \alpha.$$

The mapping E^{-1} is a diffeomorphism of the normal neighborhood $V(p)$ of the point p onto the open set $\alpha^{-1}(V_0) \subset R^m$. We shall call the corresponding local coordinates $x^1, \ldots, x^m$ at the point p [with coordinate neighborhood $V(p)$] the ***normal coordinates*** (defined by the basis $A_1, \ldots, A_m$) at the point p. Thus, by definition,

$$\mathrm{E}^{-1}(q) = (x^1(q), \ldots, x^m(q)), \quad q \in V(p).$$

From this it follows that, for an arbitrary vector $A = a^i A_i$ in the space M_p,

$$x^i(\mathrm{Exp}\, A) = a^i.$$

Since $\gamma_A(t) = \gamma_{tA}(1)$, it follows from this that

In the normal coordinates $x^1, \ldots, x^m$, every geodesic γ_A is defined in the neighborhood $V(p)$ by the linear functions

$$x^1(t) = a^1 t, \ldots, x^m(t) = a^m(t).$$

8. Whitehead's Theorem

The following theorem, first proven by Whitehead, is a considerable strengthening of the theorem proven above on the existence of normal neighborhoods:

Every point p_0 in a space M of affine connection has a neighborhood V that is a normal neighborhood of each of its points.

Since any two points $p, q \in V$ can be connected in V by one and only one geodesic (because V is a normal neighborhood, let us say, of the point p), it follows from this theorem that

Every point p_0 in a space of affine connection has a neighborhood V any two points of which can be connected in V by a unique geodesic.

To prove Whitehead's theorem, let us again consider the system of local coordinates $x^1, \ldots, x^m$ and the metric $\rho(p, q)$ defined by that system in the neighborhood $U = U_{c_1}(p_0)$ (of the point p_0) that we considered in the preceding section.

For every point $p \in U_c(p_0)$, where c is a positive number, we constructed in that section a normal neighborhood $V(p)$ and a positive number $\eta(p)$ depending continuously on p and enjoying the property that the spherical neighborhood $U_{\eta(p)}(p)$ is contained in the neighborhood $V(p)$. Since $\eta(p)$ depends continuously on p, it follows that, for arbitrary positive $\delta < c$, the greatest lower bound η_δ of the numbers $\eta(p)$ in the closure $\overline{U}_\delta(p_0)$ of the spherical δ-neighborhood $U_\delta(p_0)$ of the point p_0 is positive (since this closure, being homeomorphic to the closed Euclidean ball, is compact). Since η_δ can only increase with decreasing δ, there exists a $\delta_0 > 0$ such that $\eta_{\delta_0} > 2\delta_0$. But then, for arbitrary $\delta \leq \delta_0$ and an arbitrary point

$p \in U_\delta(p_0)$, we have

$$U_\delta(p_0) \subset U_{2\delta}(p) \subset U_{2\delta_0}(p) \subset U_{\eta_{\delta_0}}(p) \subset$$
$$\subset U_{\eta_\delta}(p) \subset U_{\eta(p)}(p) \subset V(p).$$

Thus we have proved that

There exists a positive number δ_0 *such that, for arbitrary positive* $\delta \leqslant \delta_0$ *and an arbitrary point* $p \in U_\delta(p_0)$, *the neighborhood* $U_\delta(p_0)$ *is contained in some normal neighborhood of the point* p.

From this it follows, in particular, that

Any two points p *and* q *in the neighborhood* $U_\delta(p_0)$, *can be connected by a geodesic.*

Proof: From what was proven above, there exists a normal neighborhood $V(p)$ of the point p such that $U_\delta(p_0) \subset V(p)$ and hence $q \in V(p)$. Therefore, there is a (unique) geodesic $\gamma(t)$ connecting the point p with the point q in $V(p)$.

However, in general, this geodesic does not lie in the neighborhood $U_\delta(p_0)$. To get this property also, we must assign a suitable value to the "radius" c_1 [of the coordinate neighborhood $U = U_{c_1}(p_0)$] which up to now has been completely arbitrary. Specifically, let us suppose that the number c_1 is chosen so small that, at an arbitrary point $p \in U_{c_1}(p_0)$, the matrix

$$\left(\delta_{ij} - \sum_{k=1}^{m} \Gamma_{ij}^k(p)\, x^k(p)\right)$$

is positive-definite. Since, by hypothesis, $x^1(p_0) = \ldots = x^m(p_0) = 0$, such a choice of c_1 is always possible. Let us show that

If c_1 *is chosen as indicated, the geodesic* $\gamma(t)$ *connecting the points* p *and* q *that we constructed above lies entirely in the neighborhood* $U_\delta(p_0)$.

Proof: Suppose that the parameter t on the geodesic $\gamma(t)$ is chosen in such a way that $\gamma(0) = p$, $\gamma(1) = q$. Let us set $x^i(t) = x^i(\gamma(t))$ and consider the function

$$F(t) = \sum_{i=1}^{m} (x^i(t))^2$$

on the interval [0, 1]. This function is the square of the distance $\rho(\gamma(t), p_0)$ between the points $\gamma(t)$ and p_0). Since the functions $x^i(t)$ satisfy equations (1) of section 6, we have

$$\ddot{F}(t) = 2 \sum_{k=1}^{m} ((\dot{x}^k(t))^2 + \ddot{x}^k(t)\, x^k(t)) =$$
$$= 2 \sum_{k=1}^{m} ((\dot{x}^k(t))^2 - \Gamma_{ij}^k(t)\, \dot{x}^i(t)\, \dot{x}^j(t)\, x^k(t)) =$$
$$= 2\left(\delta_{ij} - \sum_{k=1}^{m} \Gamma_{ij}^k(t)\, x^k(t)\right) \dot{x}^i(t)\, \dot{x}^j(t),$$

and, therefore,

$$\ddot{F}(t) > 0, \quad 0 \leqslant t \leqslant 1.$$

(We recall that the geodesic $\gamma(t)$ lies in the neighborhood $V(p)$ and hence in the neighborhood $U = U_{c_1}(p_0)$.)

Thus, the function $F(t)$ is a convex on the interval [0, 1] and, hence,

$$F(t) \leqslant \max(F(0), F(1)), \quad 0 \leqslant t \leqslant 1.$$

But $F(0) = \rho^2(p, p_0) < \delta^2$ and $F(1) = \rho^2(q, p_0) < \delta^2$. Therefore, $F(t) < \delta^2$; that is, $\rho(\gamma(t), p_0) < \delta$; hence $\gamma(t) \in U_\delta(p_0)$.

Thus, we have shown that, for arbitrary $\delta < \delta_0$, the neighborhood $U_\delta(p_0)$ of the point p_0 lies in some normal neighborhood $V(p)$ of any point p in it and that, in addition, every point $q \in U_\delta(p_0)$ can be connected with the point p by a geodesic lying entirely in $U_\delta(p_0)$. But then, as we know (cf. section 7), the neighborhood $U_{\delta(p_0)}$ is a normal neighborhood of the point p. This completes the proof of Whitehead's theorem.

9. Differential Connection Forms

Let p denote an arbitrary point in a space M of affine connection, let V_0 denote a normal neighborhood of the zero vector in the space M_p, and let $V(p) = \operatorname{Exp} V_0$ denote the corresponding normal neighborhood of the point p. As we know, for an arbitrary point $q \in V(p)$, there exists in $V(p)$ a unique geodesic connecting the point p with the point q. Let $\tau_q : M \to M_q$ be a parallel translation along this geodesic. To an arbitrary vector $A \in M_p$, we assign the vector $X_q \in M_q$ by setting

$$X_q = \tau_q A.$$

Obviously,

***The vectors** X_q **constitute on** $V(p)$ **a vector field** X.*

We shall say that the field X is obtained by a parallel translation of the vector A along geodesics.

Now, let $A_1, \ldots, A_m$ denote an arbitrary basis for the space M_p. Let us consider, on the neighborhood $V(p)$, the fields $X_1, \ldots, X_m$ obtained from the vectors $A_1, \ldots, A_m$ by parallel translation along geodesics. Since, for an arbitrary point $q \in V(p)$, the vectors $(X_1)_q, \ldots, (X_m)_q$ are obviously linearly independent, the fields $X_1, \ldots, X_m$ constitute a basis for the module $\mathfrak{S}^1(V(p))$ [see section 6 of Chapter 1]. We shall call this basis a ***normal basis*** (generated by the given basis $A_1, \ldots, A_m$ for the space M_p). In general, this basis is not holonomic.

Let $\bar{\omega}^1, \ldots, \bar{\omega}^m$ be a conjugate basis for the module $\mathfrak{S}_1(V(p))$, that is, let the conjugate basis be such that

$$\bar{\omega}^i(X_j) = \delta^i_j,$$

and let Γ^k_{ij} denote the coefficients of the connection ∇ in the basis $X_1, \ldots, X_m$:

$$\Gamma^k_{ij} = \bar{\omega}^k(\nabla_{X_i} X_j).$$

We set

$$\bar{\omega}^i_j = \Gamma^i_{kj}\bar{\omega}^k.$$

Let Q denote the set of all points $(t, a^1, \ldots, a^m) = (t, a)$ for $t \in R$ and $a \in R^m$ of the set R^{m+1} at which $ta^iA_i \in V_0$. Consider the mapping

$$\hat{\mathrm{E}} : Q \to V(p),$$

defined by

$$\hat{\mathrm{E}}(t, a) = \mathrm{E}(ta),$$

that is, by

$$\hat{\mathrm{E}}(t, a) = \mathrm{Exp}\,(ta^iA_i).$$

The set Q is open in R^{m+1} and hence represents a smooth manifold (of dimension $m + 1$) with local coordinates $t, a^1, \ldots, a^m$ (defined on the entire set Q). Obviously, the mapping $\hat{\mathrm{E}}$ is smooth at an arbitrary point $(t, a) \in Q$. Therefore, the linear differential forms $\hat{\mathrm{E}}^*\bar{\omega}^i$ and $\hat{\mathrm{E}}^*\bar{\omega}^i_j$ (cf. section 8, Chapter 1) are defined on Q.

Let us fix the numbers $a^1, \ldots, a^m$ and consider the curve

$$\gamma(t) = \hat{\mathrm{E}}(t, a).$$

This curve is a segment of the geodesic γ_A, where $A = a^kA_k$, so that

$$\dot{\gamma}(0) = a^kA_k.$$

Since the tangent vectors on a geodesic are parallel, it follows from this that

$$\dot{\gamma}(t) = (a^kX_k)_{\gamma(t)} \quad \text{for every } t. \tag{1}$$

On the other hand, it follows immediately from the definitions that

$$d\hat{\mathrm{E}}_{(t, a)}\left(\frac{\partial}{\partial t}\right) = d\gamma_t\left(\frac{\partial}{\partial t}\right) = \dot{\gamma}(t).$$

Thus,

$$(\hat{\mathrm{E}}^*\bar{\omega}^i)\left(\frac{\partial}{\partial t}\right) = \bar{\omega}^i(a^kX_k) = a^i.$$

This proves that the form $\hat{\mathrm{E}}^*\bar{\omega}^i$ defined on Q may be written

$$\hat{\mathrm{E}}^*\bar{\omega}^i = a^i\,dt + \omega^i,$$

where ω^i is a form composed of the differentials $da^1, \ldots, da^m$ (with coefficients depending on t).

Analogously,

$$(\hat{\mathrm{E}}^{*}\overline{\omega}_j^i)\left(\frac{\partial}{\partial t}\right)=\overline{\omega}_j^i(a^k X_k)=\Gamma_{kj}^i a^k.$$

But since the vector $(a^k X_k)_{\gamma(t)}$ is the tangent vector of the geodesic $\gamma(t)$ and since the vectors $(X_j)_{\gamma(t)}$ are parallel on $\gamma(t)$ for arbitrary j, we have

$$\nabla_{a^k X_k}(X_j)=\Gamma_{kj}^i a^k X_i=0,$$

and hence

$$\Gamma_{kj}^i a^k=0.$$

This proves that the coefficient of dt in the form $\hat{\mathrm{E}}^{*}\overline{\omega}_j^i$ is equal to 0, that is, that this form, in analogy with the forms ω^i, depends only on the differentials $da^1, \ldots, da^m$. We set

$$\omega_j^i=\hat{\mathrm{E}}^{*}\overline{\omega}_j^i.$$

We shall call these forms ω^i and ω_j^i of the differentials $da^1, \ldots, da^m$ the *connection forms.*

It is easy to see that

For $t = 0$, all the connection forms vanish (that is, all their coefficients vanish).

Proof: At the point $(0, a)$, the equation

$$d\hat{\mathrm{E}}_{(0,\,a)}\left(\frac{\partial}{\partial a^k}\right)f=\left(\frac{\partial}{\partial a^k}(f\circ\hat{\mathrm{E}})\right)_{t=0}=\frac{\partial}{\partial a^k}((f\circ\hat{\mathrm{E}})_{t=0})=0,$$

holds for an arbitrary function $f\in\mathfrak{G}(p)$ since $(f\circ\hat{\mathrm{E}})_{t=0}=0$. (We recall that the operations $\partial/\partial a^k$ as applied to the function $f\circ\hat{\mathrm{E}}$ are simply differentiations with respect to the coordinates.) Thus,

$$d\hat{\mathrm{E}}_{(0,\,a)}\left(\frac{\partial}{\partial a^k}\right)=0$$

and, hence,

$$\omega^i\left(\frac{\partial}{\partial a^k}\right)(0,\ a)=\omega_j^i\left(\frac{\partial}{\partial a^k}\right)(0,\ a)=0.$$

10. Cartan's Equations

In this section, we shall show that the connection forms satisfy a certain remarkable system of differential equations, first discovered by Cartan. To find these equations, let us first show that, for arbitrary vector fields $X, Y\in\mathfrak{G}^1(V(p))$,

$$\begin{aligned}
&X\overline{\omega}^i(Y)-Y\overline{\omega}^i(X)-\overline{\omega}^i([X,\ Y])=\\
&\qquad=\overline{\omega}^k(X)\,\overline{\omega}_k^i(Y)-\overline{\omega}^k(Y)\,\overline{\omega}_k^i(X)+K_{kl}^i\overline{\omega}^k(X)\,\overline{\omega}^l(Y),\\
&X\overline{\omega}_j^i(Y)-Y\overline{\omega}_j^i(X)-\overline{\omega}_j^i([X,\ Y])=\\
&\qquad=\overline{\omega}_j^k(X)\,\overline{\omega}_k^i(Y)-\overline{\omega}_j^k(Y)\,\overline{\omega}_k^i(X)+R_{jkl}^i\overline{\omega}^k(X)\,\overline{\omega}^l(Y),
\end{aligned}\tag{1}$$

where the K^i_{kl} and R^i_{jkl} are the components of the fields K and R of the torsion and curvature tensors of the affine connection ∇ relative to the normal basis $X_1, \ldots, X_m$.

We note that equations (1) are linear in X and Y. Therefore, it will be sufficient to verify them for basis fields $X = X_s$ and $Y = X_r$, where $s, r = 1, \ldots, m$. But by virtue of the equations

$$\bar{\omega}^i(X_j) = \delta^i_j, \quad \bar{\omega}^i_j(X_k) = \Gamma^i_{kj},$$

equations (1) for the fields $X = X_s$ and $Y = X_r$ are of the form

$$\begin{gathered} -c^i_{sr} = \Gamma^i_{rs} - \Gamma^i_{sr} + K^i_{sr}, \\ X_s\Gamma^i_{rj} - X_r\Gamma^i_{sj} - \Gamma^i_{kj}c^k_{sr} = \Gamma^k_{sj}\Gamma^i_{rk} - \Gamma^k_{rj}\Gamma^i_{sk} + R^i_{jsr}, \end{gathered} \tag{2}$$

where the c^i_{sr} are the coefficients in the decomposition of the field $[X_s, X_r]$ relative to the basis $X_1, \ldots, X_m$:

$$[X_s, X_r] = c^i_{sr}X_i.$$

Equations (2) essentially coincide with formulas (1) and (2) of section 2. Thus, we have proven equations (1) above.

Now, let X and Y be vector fields on Q. If we write equations (1) for an arbitrary point $q = \hat{\mathrm{E}}(t, a) \in V(p)$ but replace the vectors X_q and Y_q in the equations obtained by the vectors $d\hat{\mathrm{E}}_{(t,a)}(X_{(t,a)})$ and $d\hat{\mathrm{E}}_{(t,a)}(Y_{(t,a)})$ and use the formula defining the differential $d\hat{\mathrm{E}}_{(t,a)}$ (cf. section 6 Chapter 1), we immediately see that equations (1) remain valid for fields on Q if we replace the forms $\bar{\omega}^i$ and $\bar{\omega}^i_j$ in these equations with the forms $\hat{\mathrm{E}}^*\bar{\omega}^i = a^i dt + \omega^i$ and $\hat{E}^*\bar{\omega}^i_j = \omega^i_j$, respectively (and if by the functions K^i_{kl} and R^i_{jkl} we understand the functions $K^i_{kl} \circ \hat{\mathrm{E}}$ and $R^i_{jkl} \circ \hat{\mathrm{E}}$, respectively). Thus, for arbitrary vector fields $X, Y \in \mathfrak{G}^1(Q)$,

$$\begin{aligned} X(a^i dt(Y)) &+ X\omega^i(Y) - Y(a^i dt(X)) - Y\omega^i(X) - \\ &- a^i dt([X, Y]) - \omega^i([X, Y]) = \\ &= (a^k dt(X) + \omega^k(X))\omega^i_k(Y) - (a^k dt(Y) + \omega^k(Y))\omega^i_k(X) + \\ &\qquad + T^i_{kl}(a^k dt(X) + \omega^k(X))(a^l dt(Y) + \omega^l(Y)), \\ X\omega^i_j(Y) &- Y\omega^i_j(X) - \omega^i_j([X, Y]) = \\ &= \omega^k_j(X)\omega^i_k(Y) - \omega^k_j(Y)\omega^i_k(X) + \\ &\qquad + R^i_{jkl}(a^k dt(X) + \omega^k(X))(a^l dt(Y) + \omega^l(Y)). \end{aligned}$$

Let us apply these equations to the field $X = \partial/\partial t$ and to a field Y of the form $Y^i \dfrac{\partial}{\partial a^i}$, where the functions Y^i are independent of t. Since

$$dt(X) = 1, \quad \omega^i(X) = 0, \quad \omega^i_j(X) = 0, \quad dt(Y) = 0$$

and

$$[X, Y] = \frac{\partial}{\partial t}\left(Y^i \frac{\partial}{\partial a^i}\right) - Y^i \frac{\partial}{\partial a^i}\frac{\partial}{\partial t} = \frac{\partial Y^i}{\partial t}\frac{\partial}{\partial a^i} = 0,$$

in this case, we obtain the relations

$$\begin{aligned} \frac{\partial}{\partial t}\omega^i(Y) &= Ya^i + a^k\omega^i_k(Y) + K^i_{kl}a^k\omega^l(Y), \\ \frac{\partial}{\partial t}\omega^i_j(Y) &= R^i_{jkl}a^k\omega^l(Y). \end{aligned} \tag{3}$$

Obviously, for an arbitrary form $\omega = \omega_i\, da^i$, the coefficients ω_i of which depend on t, the formula

$$\frac{\partial \omega}{\partial t}(Y) = \frac{\partial}{\partial t}\omega(Y)$$

defines, for arbitrary t on the set Q_t of points $a \in R^m$ at which $(t, a) \in Q$, a form $\partial\omega/\partial t$ with coefficients $\partial\omega_i/\partial t$. Consequently, remembering that $Ya^i = da^i(Y)$, we can rewrite equations (3) in the following final form:

$$\begin{aligned} \frac{\partial \omega^i}{\partial t} &= da^i + a^k\omega^i_k + K^i_{kl}a^k\omega^l; \\ \frac{\partial \omega^i_j}{\partial t} &= R^i_{jkl}a^k\omega^l. \end{aligned} \tag{4}$$

These are the differential equations discovered by Cartan. Together with the initial conditions

$$\omega^i\big|_{t=0} = \omega^i_j\big|_{t=0} = 0$$

they enable us to determine the connection forms uniquely from the tensor fields K and R.

CHAPTER 3

RIEMANNIAN SPACES

1. Existence and uniqueness of a Riemannian connection

Let M denote an arbitrary smooth manifold of dimension m. We shall call the manifold M a *Riemannian space* if a tensor field $g(X, Y)$ of type (0, 2) is defined on it with the following two properties:

(1) for arbitrary vector fields $X, Y \in \mathfrak{G}^1(M)$,

$$g(X, Y) = g(Y, X);$$

(2) for an arbitrary point $p \in M$, the bilinear form g_p on the space M_p is positive-definite.

We shall call the tensor field g the field of the *metric tensor* of the Riemannian space M.

For an arbitrary point p in a Riemannian space M, the positive-definite bilinear form g_p defines in the tangent space M_p a scalar product $(A, B) = g_p(A, B)$ that transforms this space into a Euclidean space. This enables us, in particular, to speak of the length of vectors in M_p and of the angle between them. We shall denote the length $\sqrt{(A, A)}$ of a vector $A \in M_p$ by $|A|$.

We shall call an affine connection Δ defined on a Riemannian space M *a Riemannian connection if*

(1) its torsion tensor field K is equal to 0, that is,

$$\nabla_X Y - \nabla_Y X = [X, Y] \tag{1}$$

for arbitrary vector fields $X, Y \in \mathfrak{G}^1(M)$;

(2) a parallel translation along any curve on M is an isometric mapping of tangent spaces.

Condition (2) means that, for arbitrary points $p, q \in M$ and an arbitrary curve γ connecting the points p and q,

$$g_q = \tau g_p,$$

where τ is the parallel translation along the curve γ from the point p to the point q. It follows immediately from this that condition (2) is equivalent to satisfaction of the equation

$$\nabla_Z g = 0 \tag{2}$$

for an arbitrary vector field $Z \in 6^1(M)$.

Since the operation ∇_Z is a derivation of the algebra $O(M)$ [cf. section 5 of Chapter 2], it follows that, for arbitrary fields X, Y, $Z \in 6^1(M)$,

$$\nabla_Z (g \otimes X \otimes Y) = \nabla_Z g \otimes X \otimes Y + g \otimes \nabla_Z X \otimes Y + g \otimes X \otimes \nabla_Z Y.$$

If we now perform a complete contraction and recall that, on smooth functions, the operation ∇_Z coincides with the operation Z, we obtain

$$Z g (X, Y) = (\nabla_Z g)(X, Y) + g (\nabla_Z X, Y) + g (X, \nabla_Z Y).$$

From this it follows that equation (2) is equivalent to satisfaction of the equation

$$Z g (X, Y) = g (\nabla_Z X, Y) + g (X, \nabla_Z Y) \tag{3}$$

for any two vector fields $X, Y \in 6^1(M)$.

Furthermore, it follows from equation (3) and the definition of the field K that

$$Z g (X, Y) = g (\nabla_X Z, Y) + \\ + g (X, \nabla_Z Y) + g ([Z, X], Y) + g (K [Z, X], Y).$$

Since, at an arbitrary point $p \in M$, the bilinear form g_p is positive-definite and hence nondegenerate, it follows that condition (1) is equivalent (when condition (2) is satisfied) to the equation

$$Z g (X, Y) = g (\nabla_X Z, Y) + g (X, \nabla_Z Y) + g ([Z, X], Y). \tag{4}$$

This proves that

A connection Δ is a Riemannian connection if and only if equations (3) and (4) hold for arbitrary vector fields X, Y, $Z \in 6^1(M)$.

If we permute the symbols X, Y, and Z cyclically in equation (4), we obtain two more relations analogous to it. If we add the first of these relations to the third and then subtract the second, we obtain

$$2 g (X, \nabla_Z Y) = Z g (X, Y) - X g (Y, Z) + Y g (Z, X) + \\ + g (Z, [X, Y]) - g (X, [Y, Z]) + g (Y, [Z, X]). \tag{5}$$

Since, at an arbitrary point $p \in M$, the form g_p is nondegenerate, this relation defines the field $\nabla_Z Y$ unambiguously.

If we use formula (5) to define a vector field $\nabla_Z Y$ on an arbitrary Riemannian space M for any two vector fields $Z, Y \in 6^1(M)$, we obtain a linear mapping $\nabla_Z : 6^1(M) \to 6^1(M)$. Since the right-hand side of formula (5) is linear in Z, the mapping ∇_Z depends linearly on Z. Furthermore, by expanding the expression $2 g (X, \nabla_Z (fY))$, where

$f \in \mathfrak{G}(M)$, we see immediately that it is equal to

$$2Zf \cdot g(X, Y) + 2f \cdot g(X, \nabla_Z Y).$$

It follows from this that the rule for differentiating the product fY is valid for the mapping ∇_Z. Thus, the mapping $Z \to \nabla_Z$ is an affine connection on M. Furthermore, if we permute the symbols X and Y in formula (5) and add the resulting formula to formula (5), we obtain, after simplifying, relation (3). Analogously, if we permute the symbols X, Y, and Z cyclically in formula (5) and then add the resulting formulas to formula (5), we obtain relation (4). This means that the affine connection obtained is a Riemannian connection.

Thus, we have proven

On an arbitrary Riemannian space, there exists exactly one Riemannian connection ∇.

In what follows, we shall consider every Riemannian space as the space of an affine connection relative to this Riemannina connection ∇. Hence, all the results of Chapter 2 will be applicable to every Riemannian space.

Let $x^1, \ldots, x^m$ denote an arbitrary system of local coordinates in the Riemannian space M and let Γ^k_{ij} denote the coefficients of a connection ∇ relative to this system of local coordinates. As we know (see section 2 of Chapter 2), the connection ∇ will satisfy condition (1) of the definition of a Riemannian connection if and only if

$$\Gamma^k_{ij} = \Gamma^k_{ji}. \tag{6}$$

On the other hand, it is clear that equation (3) will be valid for arbitrary fields X, Y and Z if and only if it is valid in each coordinate neighborhood for fields of the form $\partial/\partial x^i$. But, if we set

$$X = \frac{\partial}{\partial x^i}, \; Y = \frac{\partial}{\partial x^j}, \; Z = \frac{\partial}{\partial x^k},$$

in this equation, we obtain, as one can easily see, the relation

$$\frac{\partial g_{ij}}{\partial x^k} = g_{lj}\Gamma^l_{ki} + g_{li}\Gamma^l_{kj}, \tag{7}$$

where the g_{ij} are the components of the tensor field g in the local coordinates $x^1, \ldots, x^m$. Thus, we have proven that

The connection ∇ *is a Riemannian connection if and only if its coefficients* Γ^l_{jk} *in an arbitrary system of local coordinates satisfy relations (6) and(7).*

We shall now give an important differentiation formula that follows from formula (3).

Let $\gamma(t)$ denote an arbitrary regular curve and let $X(t)$ and $Y(t)$ denote two vector fields on the curve $\gamma(t)$. In accordance with relation (3), for every elementary segment of the curve,

$$A g(X, Y) = g(\nabla_A X, Y) + g(X, \nabla_A Y),$$

where X, Y, and A are the extensions of the vector fields $X(t)$, $Y(t)$, and $\dot{\gamma}(t)$ from that elementary segment to the entire space M. But by definition,

$$g(\nabla_A X, Y)(\gamma(t)) = g_{\gamma(t)}((\nabla_A X)_{\gamma(t)}, Y_{\gamma(t)}) = \left(\frac{\nabla X}{dt}(t), Y(t)\right),$$

and, analogously,

$$g(X, \nabla_A Y) = \left(X(t), \frac{\nabla Y}{dt}(t)\right).$$

Furthermore,

$$\begin{aligned}[Ag(X, Y)](\gamma(t)) &= A_{\gamma(t)}g(X, Y) = \dot{\gamma}(t)\, g(X, Y) = \\ &= \dot{x}^k(t)\left(\frac{\partial}{\partial x^k}\right)_{\gamma(t)} g(X, Y) = \dot{x}^k(t)\left(\frac{\partial g(X, Y)}{\partial x^k}\right)_{\gamma(t)} = \\ &= \frac{d}{dt}(g(X, Y)(\gamma(t))) = \\ &\qquad = \frac{d}{dt}\left(g_{\gamma(t)}(X_{\gamma(t)}, Y_{\gamma(t)})\right) = \frac{d}{dt}(X(t), Y(t)).\end{aligned}$$

Thus, we have proven that

For arbitrary vector fields X(t) and Y(t) on a curve $\gamma(t)$,

$$\frac{d(X(t), Y(t))}{dt} = \left(\frac{\nabla X}{dt}(t), Y(t)\right) + \left(X(t), \frac{\nabla Y}{dt}(t)\right). \tag{8}$$

Remark: Formula (8) can easily be proven on the basis of equation (7) by using formula (1) of section 3, Chapter 2. A proof along these lines does not require the assumption that the curve is regular, so that

Formula (8) holds for arbitrary smooth curves $\gamma(t)$.

2. The Riemannian curvature tensor

By definition, the field K of the torsion tensor for an arbitrary Riemannian space M is equal to 0. In this connection, it is convenient to consider not the field R of the curvature tensor but the type-(0, 4) field

$$R(X, Y, Z, V) = g(R(X, Y)Z, V).$$

called the field of the Riemannian curvature tensor. It turns out that this field has the following symmetry properties:

$$R(X, Y, Z, V) = -R(Y, X, Z, V); \tag{1}$$

$$R(X, Y, Z, V) = -R(X, Y, V, Z); \tag{2}$$

$$R(X, Y, Z, V) = R(Z, V, X, Y). \tag{3}$$

Property (1) follows immediately from the relation $R(X, Y) = -R(Y, X)$. To prove property (2), it suffices to prove it at an arbitrary point $p \in M$ and, to do this, it suffices to show that, for any three vectors $A, B, C \in M_p$,

$$R_p(A, B, C, C) = 0.$$

Let $V(p)$ denote a normal neighborhood of the point p and let Z denote the vector field on $V(p)$ obtained by parallel translation of the vector C along geodesics (see beginning of section 9, Chapter 2). In accordance with formula (5) of section 1,

$$2g(Z, \nabla_X Z) = Xg(Z, Z)$$

for an arbitrary vector field X on $V(p)$. But since the field Z was obtained by a parallel translation that preserves the field g, we have

$$g_q(Z_q, Z_q) = g_p(C, C),$$

that is, for an arbitrary point $q \in V(p)$,

$$g(Z, Z) = \text{const} \quad \text{on} \quad V(p)$$

and, therefore,

$$Xg(Z, Z) = 0.$$

Thus,

$$g(Z, \nabla_X Z) = 0.$$

From this and from formula (3) of section 1, it follows that, for arbitrary fields $X, Y \in \mathfrak{G}^1(V(p))$,

$$R(X, Y, Z, Z) = g(R(X, Y)Z, Z) = g(\nabla_X \nabla_Y Z, Z) - \\ - g(\nabla_Y \nabla_X Z, Z) = Xg(\nabla_Y Z, Z) - Yg(\nabla_X Z, Z) = 0.$$

Consequently, by setting $X_p = A$ and $Y_p = B$, we obtain $R_p(A, B, C, C) = 0$. This completes the proof of property (2).

To prove property (3), we note that, on the basis of Bianchi's identity (cf. section 2 of Chapter 2), an analogous identity holds for the Riemannian curvature tensor:

$$R(X, Y, Z, V) + R(Y, Z, X, V) + R(Z, X, Y, V) = 0. \qquad (4)$$

If we interchange the symbol V in this equation successively with the symbols X, Y and Z, combine the three resulting equations with equation (4), and keep properties (1) and (2) in mind, we obtain

$$R(V, Y, Z, X) + R(V, Z, X, Y) = 0.$$

Analogously,

$$R(Y, X, V, Z) + R(Y, V, Z, X) = 0.$$

If we add the last two relations and keep property (2) in mind, we obtain the equation

$$R(Y, X, V, Z) + R(V, Z, X, Y) = 0,$$

which is equivalent to equation (3). This completes the proof of property (3).

3. Differential connection forms and the metric tensor

Let us now look at the connection forms ω^i and ω^i_j corresponding to a Riemannian manifold M. Since $K = 0$, Cartan's equations for them take the form

$$\begin{aligned} &\frac{\partial \omega^i}{\partial t} = da^i + a^k \omega^i_k, \quad \omega^i |_{t=0} = 0, \\ &\frac{\partial \omega^i_j}{\partial t} = R^i_{jkl} a^k \omega^l, \quad \omega^i_j |_{t=0} = 0. \end{aligned} \tag{1}$$

We have constructed the differential forms ω^i and ω^i_j by starting with some normal basis $X_1, \ldots, X_m$ in a normal neighborhood $V(p)$ of the point p. In the case of a Riemannian space, we shall also make the assumption that this basis is orthonormal, that is, that the relations

$$g(X_i, X_j) = \delta_{ij} \tag{2}$$

hold at the point p and hence (because of the normality of the basis and the invariance of the tensor field g under parallel translation) at an arbitrary point $q \in V(p)$. Because of this assumption, the functions $g(X_i, X_j)$ are constant on $V(p)$ and therefore

$$X_k g(X_i, X_j) = 0.$$

It follows from this on the basis of equation (3) of section 1 that

$$g(\nabla_{X_k} X_i, X_j) + g(X_i, \nabla_{X_k} X_j) = 0.$$

But, by definition,

$$g(\nabla_{X_k} X_i, X_j) = g(\Gamma^h_{ki} X_h, X_j) = \Gamma^h_{ki} \delta_{hj} = \Gamma^j_{ki},$$

and, analogously,

$$g(X_i, \nabla_{X_k} X_j) = \Gamma^i_{kj}.$$

Thus,

$$\Gamma^j_{ki} + \Gamma^i_{kj} = 0,$$

that is,

$$\bar{\omega}^j_i + \bar{\omega}^i_j = 0,$$

and, hence,

$$\omega_i^j + \omega_j^i = 0. \tag{3}$$

Furthermore, by definition, $\bar{\omega}^i(X_j) = \delta_j^i$. Consequently, by computing the value of the form $\bar{\omega}^i \otimes \bar{\omega}^i$ from the fields X_j and X_k, we obtain

$$(\bar{\omega}^i \otimes \bar{\omega}^i)(X_j, X_k) = \bar{\omega}^i(X_j)\,\bar{\omega}^i(X_k) = \delta_j^i \delta_k^i.$$

It follows immediately from this that

$$g = \sum_i \bar{\omega}^i \otimes \bar{\omega}^i$$

(since, for all fields, both sides of this equation assume the same values). Thus,

$$\hat{E}^* g = \sum_i \hat{E}^* \bar{\omega}^i \otimes \hat{E}^* \bar{\omega}^i.$$

Remembering that

$$\hat{E}^* \bar{\omega}^i = a^i\,dt + \omega^i,$$

we then obtain

$$\begin{aligned} \hat{E}^* g = \Big[\sum_i (a^i)^2\Big] dt \otimes dt + \Big(\sum_i a^i \omega^i\Big) \otimes dt + \\ + dt \otimes \Big(\sum_i a^i \omega^i\Big) + \sum_i \omega^i \otimes \omega^i. \end{aligned} \tag{4}$$

Let us now consider the subset Q_0 of the set Q consisting of points $(t, a) \in Q$ at which

$$(a^1)^2 + \ldots + (a^m)^2 = 1.$$

This set, as the regular level surface $[\varphi = 1]$ of the function

$$\varphi(t, a) = \sum_i (a^i)^2,$$

is a smooth manifold (cf. section 14, Chapter 1). Let $\iota\colon\ Q_0 \to Q$ denote the embedding mapping. We have

$$\iota^*\Big(\sum_i a^i \omega^i\Big) = 0. \tag{5}$$

This is true because, on the basis of Cartan's equations and equations (3),

$$\frac{\partial}{\partial t}\Big(\sum_i a^i \omega^i\Big) = \sum_i a^i\,da^i = d\Big(\sum_i (a^i)^2\Big) = d\varphi$$

and, hence,

$$\frac{\partial}{\partial t}\left(\iota^* \sum_i a^i \omega^i\right) = \iota^* \frac{\partial}{\partial t}\left(\sum_i a^i \omega^i\right) = \iota^* \, d\varphi = d\,(\varphi \circ \iota) = 0,$$

since $\varphi \circ \iota = 1$. Thus, the form

$$\iota^* \sum_i a^i \omega^i$$

is independent of t. But, by virtue of the initial conditions $\omega^i|_{t=0} = 0$, this form vanishes at $t = 0$. Consequently, it vanishes for arbitrary t.

Now, let H denote the restriction $\hat{\mathrm{E}}|_{Q_0} = \hat{\mathrm{E}} \circ \iota$ of the mapping $\hat{\mathrm{E}}$ to Q_0. Then,

$$\mathrm{H}^* g = \iota^*(\hat{\mathrm{E}}^* g).$$

On the basis of formulas (4) and (5), it follows from this that

$$\mathrm{H}^* g = dt \otimes dt + \sum_i \omega^i \otimes \omega^i, \tag{6}$$

where t and ω^i denote, respectively, the function $t|_{Q_0}$ and the form $\iota^* \omega^i$.

We shall often have occasion to use this important relationship. In this connection, it is worth noting that

The mapping H *is a diffeomorphism of the set* Q_0^+ *of points* $(t, a) \in Q_0$ *at which* $t > 0$ *onto the "punctured neighborhood"* $V(p) \setminus p$ *of the point* p.

Here, if $q = \operatorname{Exp} A$, where $q \in V(p) \setminus p$ and $A \in V_0 \setminus 0$, then

$$q = \mathrm{H}(|A|, a), \tag{7}$$

where $a = (a^1, \ldots, a^m)$ is the set of coordinates of the unit vector $A/|A|$ relative to the basis $A_1, \ldots, A_m$ by means of which the mapping H (or, more precisely, the corresponding mapping E [cf. section 7 of Chapter 2]) is constructed.

4. Arc length

Let $\gamma(s)$, for $\alpha \leqslant s \leqslant \beta$, be an arbitrary piecewise-smooth curve (or, more precisely, a segment of such a curve) in a Riemannian space M. We shall call the number

$$J(\gamma) = \int_a^b |\dot{\gamma}(s)|\, ds$$

the *length* of the curve γ. (The fact that the integrand may not be defined at certain points in the interval $[\alpha, \beta]$ does not keep the integral from being meaningful.)

If the curve $\gamma(s)$, for $\alpha \leqslant s \leqslant \beta$, lies entirely in some normal neighborhood $V(p)$ and does not pass through the point p, it is convenient, when evaluating its length, to shift over to the corresponding set Q_0^+, that is, to consider the curve

$$\gamma_0 = \mathrm{H}^{-1} \circ \gamma$$

on Q_0^+, where H is the diffeomorphism $Q_0^+ \to V(p) \setminus p$ considered at the end of section 3. Since $\gamma = \mathrm{H} \circ \gamma_0$, we have

$$\dot{\gamma}(s) = d\mathrm{H}_{\gamma_0(s)}(\dot{\gamma}_0(s)),$$

and, hence,

$$|\dot{\gamma}(s)|^2 = g_{\gamma(s)}(\dot{\gamma}(s), \dot{\gamma}(s)) = (\mathrm{H}^*g)_{\gamma_0(s)}(\dot{\gamma}_0(s), \dot{\gamma}_0(s)).$$

On the basis of formula (6) of section 3, it follows immediately from this that

$$|\dot{\gamma}(s)|^2 = [dt(\dot{\gamma}_0(s))]^2 + \sum_i [\omega^i(\dot{\gamma}_0(s))]^2.$$

On the other hand, if we set

$$\gamma_0(s) = (t(s), a(s)), \qquad \alpha \leqslant s \leqslant \beta,$$

where $t(s)$ is a positive real function of the parameter s and $a(s)$ is a piecewise-smooth (though not, in general, regular [even if the curve $\gamma(s)$ is regular]) curve on the sphere S:

$$(a^1)^2 + \dots + (a^m)^2 = 1,$$

we immediately obtain

$$\dot{\gamma}_0(s) = d\gamma_0\left(\frac{\partial}{\partial s}\right) = \dot{t}(s)\frac{\partial}{\partial t} + \dot{a}(s),$$

where $t(s)$ is the derivative of the function $t(s)$ with respect to s and

$$\dot{a}(s) = da\left(\frac{\partial}{\partial s}\right)$$

is the vector tangent to the curve $a(s)$ on the sphere S. In other words,

$$dt(\dot{\gamma}_0(s)) = \dot{t}(s), \quad \omega^i(\dot{\gamma}_0(s)) = \omega^i(\dot{a}(s)).$$

From this we obtain the following formula for the number $|\dot{\gamma}(s)|$:

$$|\dot{\gamma}(s)|^2 = [\dot{t}(s)]^2 + \sum_i [\omega^i(\dot{a}(s))]^2.$$

Thus, for the length $J(\gamma)$ of the curve γ we have the formula

$$J(\gamma) = \int_\alpha^\beta \sqrt{[\dot{t}(s)]^2 + \sum_i [\omega^i(\dot{a}(s))]^2}\, ds. \tag{1}$$

We shall say that two piecewise-smooth curves $\gamma_1(t)$ for $\alpha_1 \leqslant t \leqslant \beta_1$, and $\gamma_2(s)$, for $\alpha_2 \leqslant s \leqslant \beta_2$, are equivalent if there exists on the interval $[\alpha_2, \beta_2]$ a monotonic smooth (except at points of discontinuity) function $t(s)$ such that $t(\alpha_2) = \alpha_1$, $t(\beta_2) = \beta_1$, and $\gamma_2(s) = \gamma_1(t(s))$ for all $s \in [\alpha_2, \beta_2]$. For points $s_0 \in [\alpha_2, \beta_2]$ at which the function $t(s)$ has a discontinuity of the first kind (a saltus), this equation means that

$$\gamma_2(s) = \gamma_1(t)$$

for arbitrary t satisfying the inequalities

$$t(s-0) \leqslant t \leqslant t(s+0).$$

Obviously, equivalence of curves, as we have defined it, is reflexive, symmetric, and transitive, so that the set of all curves can be decomposed into classes of equivalent curves.

Since $\dot{\gamma}_2(s) = \dot{\gamma}_1(t(s))\dot{t}(s)$ for almost all s, we obviously have

Equivalent curves have the same length.

We note that it is quite possible for a nonregular curve to be equivalent to a regular curve.

Let $p_0 = \gamma(t_0)$ be some fixed point on the curve $\gamma(t)$. We define a function $s(t)$ by

$$s(t) = \int_{t_0}^{t} |\dot{\gamma}(t)|\,dt.$$

Thus,

$$s(t) = J(\gamma|_{[t_0, t]}) \quad \text{if} \quad t \geqslant t_0,$$

and

$$s(t) = -J(\gamma|_{[t, t_0]}) \quad \text{if} \quad t \leqslant t_0.$$

Obviously, the function $s(t)$ is monotonic and continuous. It is constant if and only if the curve $\gamma(t)$ is degenerate. Its inverse function $t(s)$ obviously has the property that, for an arbitrary point s at which it has a saltus,

$$\gamma(t(s-0)) = \gamma(t(s+0))$$

(since the curve $\gamma|_{[t(s-0),\, t(s+0)]}$ is degenerate). Therefore, if the curve γ is nondegenerate, the piecewise-smooth curve

$$\gamma^*(s) = \gamma(t(s)),$$

equivalent to the curve γ is defined. This curve γ^* has the property that the length of an arbitrary segment $\gamma^*|_{[s_0, s_1]}$ of it is equal to the corresponding increment in the parameter s:

$$J(\gamma^*|_{[s_0, s_1]}) = s_1 - s_0.$$

For curves with this property, we shall refer to the parameter s as the *arc length* (measured from the point p_0 along the curve). Thus,

An arbitrary nondegenerate piecewise-smooth curve is equivalent to a curve the parameter of which is arc length.

In connection with this, it is useful to note that

Arc length is a parameter on a smooth curve $\gamma(s)$ *if and only if, for arbitrary s, the vector* $\dot{\gamma}(s)$ *is a unit vector, that is, if* $|\dot{\gamma}(s)| = 1$.

Therefore, in particular,

An arbitrary smooth curve for which arc length serves as parameter is a regular curve.

Consequently,

Every nondegenerate piecewise-smooth curve is equivalent to some piecewise-regular curve.

For geodesics, the condition $|\dot{\gamma}(s)| = 1$ needs to be verified only at a single point (since, on the geodesic $\gamma(s)$, the vectors $\dot{\gamma}(s)$ are obtained from each other by a parallel translation and hence have the same length). In particular,

Arc length is a parameter on a maximal geodesic γ_A *if and only if* $|A| = 1$.

5. The interior metric

Let M denote a connected Riemannian space; that is, any two points in M can be connected, in M, by a piecewise-smooth curve.

Remark: It is easy to see that a Riemannian space M (or, more generally, any smooth manifold) is connected if and only if it is connected as a topological space, that is, if it is not the union of two nonempty closed (or open) disjoint sets. On the other hand, it can be shown that any two points in a connected manifold can be connected by a regular curve. We state these facts without proof since we shall not have occasion to use them.

Consider the greatest lower bound of lengths of all piecewise-smooth (or, equivalently, all piecewise-regular) curves γ connecting two arbitrary points p and q of a connected Riemannian space M. We shall call this bound the distance between the points p and q and denote it by $\rho(p, q)$:

$$\rho(p, q) = \inf_{\gamma} J(\gamma).$$

It turns out that

The distance $\rho(p, q)$ *transforms the Riemannian space M into a metric space; that is, this distance is symmetric:*

$$\rho(p, q) = \rho(q, p), \qquad p, q \in M,$$

it satisfies the triangle inequality:

$$\rho(p, r) \leqslant \rho(p, q) + \rho(q, r), \quad p, q, r \in M,$$

and it is nondegenerate; that is, $\rho(p, q) = 0$ *if and only if* $p = q$.

The first two properties are obvious. To prove the third, we need only show that

$$\rho(p, q) > 0, \quad \text{if} \quad p \neq q, \tag{1}$$

since it is clear that $\rho(p, q) = 0$ when $p = q$.

Let us consider first the case in which the point q belongs to some normal neighborhood $V(p)$ of the point p. Let $\gamma(s)$, where $\alpha \leqslant s \leqslant \beta$ denote any piecewise-smooth curve in $V(p)$ connecting the point p with the point q and let s_0 denote the greatest value of the parameter s such that $\gamma(s_0) = p$ (obviously, such an s_0 always exists). Consider the segment $\gamma' = \gamma|_{[s_0, \beta]}$ of the curve γ. Obviously,

$$J(\gamma) \geqslant J(\gamma').$$

On the other hand, for arbitrary positive ε, the segment $\gamma'_\varepsilon = \gamma|_{[s_0+\varepsilon, \beta]}$ of the curve γ' lies in $V(p) \setminus p$. Therefore, formula (1) of section 4 is valid for its length. From this it follows that

$$J(\gamma'_\varepsilon) \geqslant \left| \int_{s_0+\varepsilon}^{\beta} \dot{t}(s)\, ds \right| = |t(\beta) - t(s_0 + \varepsilon)|.$$

Since, obviously, $J(\gamma'_\varepsilon) \to J(\gamma')$ and $t(s_0 + \varepsilon) \to t(s_0) = 0$ as $\varepsilon \to 0$, it follows from this and the preceding inequality that

$$J(\gamma) \geqslant t_q, \tag{2}$$

where $t_q = t(\beta)$ is the value of the coordinate $t > 0$ for which $\mathrm{H}(t_q, a) = q$.

By definition, the normal neighborhood $V(p)$ is the image, under the mapping Exp, of some normal neighborhood V_0 of the zero vector of the space M_p. If this neighborhood V_0 consists of all vectors $A \in M_p$ such that $|A| < \delta$, we shall call the corresponding neighborhood $V(p)$ the ***normal spherical*** δ***-neighborhood*** of the point p and shall denote it by $V_\delta(p)$. It follows immediately from inequality (2) and formula (7) of section 3 that, if the point q belongs to the boundary of the neighborhood $V_\delta(p)$, then

$$J(\gamma) \geqslant \delta \tag{3}$$

for an arbitrary curve γ in $V_\delta(p)$ that connects the point p with the point q. Since an arbitrary curve in M issuing from the point p and not lying entirely in $V_\delta(p)$ must necessarily intersect the boundary of the neighborhood $V_\delta(p)$, it follows that inequality (3) is valid both for an arbitrary curve connecting the point p with a point $q \in V_\delta(p)$ but not lying entirely in $V_\delta(p)$ and for an arbitrary curve in M that connects the point p with any point $q \notin V_\delta(p)$.

Thus, we have shown that either inequality (2) or inequality (3) is valid for any point q in the space M and any curve γ connecting the point p with the point q. Consequently,

$$\rho(p, q) = \inf_{\gamma} J(\gamma) \geqslant \delta_q,$$

where $\delta_q = t_q$ if $q \in V_\delta(p)$ and $\delta_q = \delta$ if $q \notin V_\delta(p)$. Since $t_q > 0$ if $q \neq p$, we have completed the proof that the distance which we have defined is nondegenerate.

We shall call the metric $\rho(p, q)$ that we have constructed the *interior metric* of the Riemannian space M.

6. Minimizing curves

We shall call a piecewise-smooth curve γ connecting points p and q a *minimizing curve* if its length is the least of the lengths of all curves on M connecting these two points, that is, if

$$J(\gamma) = \rho(p, q).$$

In general, for arbitrary points p and q, there may not exist such a minimizing curve connecting them or, on the other hand, there may be several such. However,

If the points p and q are sufficiently close to each other, that is, if the point q lies in some normal spherical δ-neighborhood $V_\delta(p)$ of the point p, there exists exactly one (up to equivalence) minimizing curve connecting p and q. This minimizing curve is the geodesic γ_q connecting the point p in $V_\delta(p)$ with the point q.

Proof: If we set $q = \operatorname{Exp} A$, where $|A| < \delta$, we see that the geodesic γ_q is a segment $\gamma_A|_{[0,1]}$ of the maximal geodesic γ_A defined by the relations

$$\gamma_A(0) = p, \quad \dot{\gamma}_A(0) = A.$$

Consequently,

$$J(\gamma_q) = \int_0^1 |\dot{\gamma}_A(t)|\, dt = \int_0^1 |A|\, dt = |A|. \tag{1}$$

Since $|A| = t_q$, where, just as above, t_q denotes the value of the coordinate t at which $\mathrm{H}(t_q, a) = q$, it follows, on the basis of inequality (2) of section 5, that

$$J(\gamma) \geqslant J(\gamma_q)$$

for an arbitrary curve γ connecting the points p and q in $V_\delta(p)$. Furthermore, since $|A| < \delta$, this inequality is, by virtue of inequality (3) of section 5, valid for an arbitrary curve γ connecting the points p and q in M. This proves that the geodesic γ_q is indeed a minimizing curve in M connecting the points p and q.

Now, let us prove that this minimizing curve is unique. Let $\gamma(s)$, where $\alpha \leqslant s \leqslant \beta$, denote an arbitrary minimizing curve in M connecting the points p and q. Without loss of generality, we can assume that it is piecewise-regular. Since

$$J(\gamma) = J(\gamma_q) = |A| < \delta,$$

it follows on the basis of inequality (3) of section 5 that the minimizing curve γ lies entirely in the neighborhood $V_\delta(p)$. Therefore, (conditional) inequality (2) of section 5 is valid for it. But since the curve γ is a minimizing curve and $J(\gamma) = |A| = t_q$, this conditional inequality becomes equality. On the other hand, it follows immediately from its proof that equality holds if and only if the following three conditions are satisfied: (1) the number α is the only value of the parameter $s \in [\alpha, \beta]$ at which $\gamma(s) = p$; (2) the function $\omega^i(\dot{a}(s)) = 0$ vanishes for all i and all $s \in [\alpha, \beta]$ (that is, $\dot{a}(s) = 0$ and hence $a(s) = \text{const}$); and (3) the function $\dot{t}(s)$ is nonnegative. But in this case, by virtue of the piecewise regularity of the curve γ, the function $\dot{t}(s)$ can vanish only at a finite number of points. Therefore, the function $t(s)$ has a continuous inverse function $s(t)$. Obviously, the curve $\gamma(s(t))$, which is equivalent to the curve $\gamma(t)$, coincides with the geodesic γ_q. This completes the proof of the assertion made above.

We used the assumption that the neighborhood $V_\delta(p)$ is a spherical neighborhood only to compare the length of the geodesic γ_q with the lengths of the curves γ that do not lie entirely in the neighborhood $V_\delta(p)$. Therefore, for an arbitrary normal neighborhood $V(p)$ of the point p,

The geodesic γ_q connecting the point p with the point q in some normal neighborhood $V(p)$ of the point p is the unique minimizing curve in $V(p)$ connecting the points p and q.

Let us note now that formula (1) and inequality (3) of section 5 immediately imply that

A point $q \in M$ belongs to a normal spherical δ-neighborhood of the point p if and only if $\rho(p, q) < \delta$, that is, if and only if the point q belongs to the spherical δ-neighborhood $U_\delta(p)$ of the point p in the metric ρ.

In other words,

$$V_\delta(p) = U_\delta(p)$$

for every positive δ such that the neighborhood $V_\delta(p)$ is defined.

Since the neighborhoods $V_\delta(p)$ obviously form a fundamental system of neighborhoods at the point p, it follows that

The neighborhoods $U_\delta(p)$ constitute a fundamental system of neighborhoods at the point p.

This means that

The metric ρ is compatible with the topology of the space M.

On the basis of a well-known theorem of P. S. Aleksandrov (cf. preface) which states that an arbitrary connected locally compact metric space is separable (satisfies the second axiom of countability), we see that

An arbitrary connected Riemannian space is separable.

Remark: Thus, not every (connected) smooth manifold can be a Riemannian space: For this it is necessary that the space satisfy the second axiom of countability. It turns out that this necessary condition is also sufficient; that is,

On every connected smooth manifold M that satisfies the second axiom of countability, there exists a tensor field g of type $(0, 2)$ that makes this manifold into a Riemannian space.

Proof: The fact that the manifold M satisfies the second axiom of countability implies that it can be covered by a countable system U_n, where $n = 1, 2, \ldots$, of coordinate neighborhoods. These neighborhoods can be chosen in such a way that the closure $\bar{U}_n$ of each of them will be compact and the local coordinates $x_n^1, \ldots, x_n^m$ defined in the neighborhood U_n will also be local coordinates in some larger neighborhood containing the set $\bar{U}_n$.

Let us choose such a system of coordinate neighborhoods U_n and consider the sets

$$A_n = \bigcup_{k=1}^{n} \bar{U}_n, \quad n = 1, 2, \ldots .$$

Each of these sets is compact and their union is the entire manifold M.

Let us consider the case $n = 1$. Let us construct on the set $A_1 = \bar{U}_1$ a tensor field g, by declaring that its components

$$g_{ij} = g\left(\frac{\partial}{\partial x_1^i}, \frac{\partial}{\partial x_1^j}\right), \quad i, j = 1, \ldots, m,$$

relative to the basis $\frac{\partial}{\partial x_1^1}, \ldots, \frac{\partial}{\partial x_1^m}$ are the numbers δ_{ij}. Obviously, the field g is now completely defined on A_1 and satisfies all the necessary conditions (symmetry and positive-definiteness).

Suppose now that the field g is constructed on the set A_{n-1}, where $n > 1$. If $A_{n-1} \cap \bar{U}_n = \varnothing$, then, by constructing on $\bar{U}_n$ a field g analogous to the one constructed on $\bar{U}_1$, we obviously obtain the required field on all $\bar{A}_n$. Suppose that $A_{n-1} \cap \bar{U}_n \neq \varnothing$. Since the field g is constructed on A_{n-1}, it is constructed, in particular, on the set $A_{n-1} \cap \bar{U}_n$, and hence the numbers

$$g_{ij}(p) = g\left(\frac{\partial}{\partial x_n^i}, \frac{\partial}{\partial x_n^j}\right), \quad i, j = 1, \ldots, m$$

are defined for an arbitrary point $p \in A_{n-1} \cap \bar{U}_n$. The set $m(m+1)/2$ of smooth functions $g_{ij}(p) = g_{ji}(p)$ can be regarded as a smooth mapping of the compact set $A_{n-1} \cap \bar{U}_n$ into the set D of all symmetric positive-definite $m \times m$ matrices. The space of all symmetric $m \times m$ matrices is isomorphic to the $[m(m+1)/2]$-dimensional Euclidean space $R^{\frac{m(m+1)}{2}}$, and, as one can easily see, the set D is a convex open subset of this space. Consequently, the set D has property E (cf. section 3 of Chapter 1). Therefore, the smooth mapping $A_{n-1} \cap \bar{U}_n \to D$ constructed above is the restriction of some smooth mapping $M \to D$; that is, on the manifold M, there exist $m(m+1)/2$ smooth functions $g_{ij}(p) = g_{ji}(p)$ that coincide on $A_{n-1} \cap \bar{U}_n$ with the functions $g_{ij}(p)$ constructed above, where $p \in A_{n-1} \cap \bar{U}_n$, and that have the property that, for an arbitrary point $p \in M$, the matrix $(g_{ij}(p))$ is symmetric and positive-definite. Obviously, if, for an arbitrary point $p \in \bar{U}$, we choose the numbers $g_{ij}(p)$ as the components relative to the basis

$$\frac{\partial}{\partial x_n^1}, \ldots, \frac{\partial}{\partial x_n^m}$$

of some tensor field g, we shall obtain on $\bar{U}_n$ a tensor field that coincides on $A_{n-1} \cap \bar{U}_n$ with the field g already constructed on A_{n-1}. In other words, we have extended the field g from A_{n-1} to A_n.

Thus, the field g is constructed by induction on an arbitrary set A_n and hence on the entire manifold M.

7. Normal convex neighborhoods

In accordance with Whitehead's theorem, which we proved in section 8 of Chapter 2, every point p_0 of the space M has a neighborhood $V(p_0)$ that is a normal neighborhood of every point in it. Furthermore, in accordance with the proof given in that section, we may take for the neighborhood $V(p_0)$ the spherical neighborhood $U_\delta(p_0)$, where $\delta < \delta_0$, of the point p_0 with respect to some metric defined in a neighborhood of the point p_0 and depending on the system of local coordinates $x^1, \ldots, x^m$ at the point p_0. It follows immediately from the definition of this metric that, if we take for $x^1, \ldots, x^m$, the normal coordinates at the point p_0, then the neighborhood $U_\delta(p_0)$ will be the image under the mapping Exp of a spherical neighborhood $|A| < \delta$ of the vector $0 \in M_p$; that is, it will be the normal spherical neighborhood $V_\delta(p_0)$. Thus,

For sufficiently small $\delta > 0$, the normal spherical δ-neighborhood $V_\delta(p_0)$ is a normal neighborhood of each of its points.

We shall call the spherical neighborhood $V_\delta(p_0)$ a *normal convex neighborhood* if it is a normal neighborhood of each of its points and if, for arbitrary points $p, q \in V_\delta(p_0)$, the geodesic γ_{pq} connecting in $V_\delta(p_0)$ the points p and q is the unique minimizing curve connecting the points p and q. It turns out that

For sufficiently small positive δ, the neighborhood $V_\delta(p_0)$ is a normal convex neighborhood.

Proof: Let δ_0 denote the least upper bound of numbers δ such that the neighborhood $V_\delta(p_0)$ is a normal neighborhood of each of its points and suppose that $\delta < \delta_0/4$. As we know, the geodesic γ_{pq}, which is contained in $V_\delta(p_0)$, is the unique curve in $V_{\delta_0}(p_0) = V(p)$ of length $\rho(p, q)$ that connects the points p and q. On the other hand, an arbitrary curve in M connecting the points p and q and not lying in $V_{\delta_0}(p_0)$ is obviously of length $> 3\delta$ and hence cannot be a minimizing curve since

$$\rho(p, q) \leqslant \rho(p, p_0) + \rho(p_0, q) \leqslant 2\delta.$$

From the assertion just proven, it follows immediately that

For an arbitrary compact Riemannian space M, there exists a number $d = d_M$ *such that arbitrary points p,* $p, q \in M$ *at a distance* $\rho(p, q)$ *from each other less than d can be joined in M by a unique minimizing curve.*

Proof: The compactness of the space implies the existence of points $p_1, \ldots, p_r \in M$ and positive numbers $\delta_1, \ldots, \delta_r$ such that the neighborhoods $V_{2\delta_1}(p_1), \ldots, V_{2\delta_r}(p_r)$ are normal convex neighborhoods of the points $p_1, \ldots, p_r$ respectively and the neighborhoods $V_{\delta_1}(p_1), \ldots, V_{\delta_r}(p_r)$ cover the entire space M. But then, an arbitrary number $d < \min(\delta_1, \ldots, \delta_r)$ will obviously have the required property.

Furthermore, it is easy to see that

On an arbitrary Riemannian space, every minimizing curve is (up to equivalence) a geodesic.

Proof: Every point on a minimizing curve belongs to some normal convex neighborhood, and the segment of the minimizing curve that lies in that neighborhood is, as we know, a geodesic (since every segment of a minimizing curve is a minimizing curve). Consequently, in a neighborhood of each of its points, the minimizing curve that we are considering is a geodesic. Therefore, the entire minimizing curve is itself a geodesic. It follows from this assertion, in particular, that

Under suitable choice of the parameter t, every minimizing curve (connecting two distinct points) is a regular curve.

Remark: It should not be supposed that the metric used in the proof of Whitehead's theorem (and constructed with the aid of normal coordinates at the point p_0) coincides with the metric ρ (although, as we have seen, the spherical neighborhoods of the point p_0 coincide in the two metrics). These metrics are equivalent only at the point p_0; that is, the ratio of the distances in the two metrics between arbitrary points p an q approaches one as $(p, q) \to (p_0, p_0)$. We state this fact without proof.

8. A lemma on convergence

Let p_0 denote an arbitrary point in a Riemannian space M and let $V = V_\delta(p_0)$ denote some normal convex neighborhood of it. Then, for an arbitrary point $p \in V$, for an arbitrary vector $B \in M_p$, and for an arbitrary sufficiently small positive number σ, the segment of length σ on the maximal geodesic γ_B issuing from the point p lies entirely in V. We denote this segment by $\gamma_{p, B, \sigma}$, and assume that the vector B is a unit vector, so that the parameter on this segment of the geodesic is the arc length s.

Let $x^1, \ldots, x^m$ denote an arbitrary system of local coordinates defined in the neighborhood V and let $x^i(s)$, where $0 \leqslant s \leqslant \sigma$, denote functions defining the segment $\gamma_{p, B, \sigma}$ parametrically in this system of coordinates and hence satisfying on the interval $[0, \sigma]$ the differential equations of the geodesics

$$\ddot{x}^i + \Gamma^i_{jk}(x^1, \ldots, x^m)\dot{x}^j\dot{x}^k = 0 \tag{1}$$

with initial conditions

$$x^i(0) = a^i, \quad \dot{x}^i(0) = b^i,$$

where

$$a^i = x^i(p), \quad b^i = Bx^i, \quad i = 1, \ldots, m.$$

Let us use the theorem on the dependence of the solutions of differential equations on the initial conditions. In accordance with this theorem, there exist on the interval $[0, \sigma]$ smooth functions

$$\bar{x}^i(s) = x^i(s; \bar{a}^1, \ldots, \bar{a}^m; \bar{b}^1, \ldots, \bar{b}^m),$$

(depending smoothly on the parameters $\bar{a}^1, \ldots, \bar{a}^m, \bar{b}^1, \ldots, \bar{b}^m$) such that, for $\bar{a}^i$ sufficiently close to a^i and $\bar{b}^i$ sufficiently close to b^i, the functions $\bar{x}^i(s)$ satisfy the system (1) and the initial conditions

$$\bar{x}^i(0) = \bar{a}^i, \quad \dot{\bar{x}}^i(0) = \bar{b}^i.$$

Since, for arbitrary $s \in [0, \sigma]$, the numbers $\bar{x}^i(s)$ are close to the numbers $x^i(s)$, there exists in V a point $\bar{\gamma}(s)$ such that $x^i(\bar{\gamma}(s)) = \bar{x}^i(s)$. Since the functions $\bar{x}^i(s)$ satisfy the system (1), the mapping $s \to \bar{\gamma}(s)$ of the interval $[0, \sigma]$ into the space M is a geodesic. It issues from the point $\bar{p}$ with coordinates $x^i(\bar{p}) = \bar{a}^i$ and has a tangent vector $\bar{B}$ at that point with coordinates $\bar{B}x^i = \bar{b}^i$.

Thus, for an arbitrary point $\bar{p}$ sufficiently close to the point p and for an arbitrary vector $\bar{B} \in M_{\bar{p}}$ sufficiently close to the vector $B \in M_p$ (in the sense that its coordinates $\bar{b}^i = \bar{B}x^i$ relative to the basis

$$\left(\frac{\partial}{\partial x^1}\right)_{\bar{p}}, \ldots, \left(\frac{\partial}{\partial x^m}\right)_{\bar{p}}$$

are sufficiently close to the coordinates $b^i = Bx^i$ of the vector B relative to the basis

$$\left(\frac{\partial}{\partial x^1}\right)_p, \ldots, \left(\frac{\partial}{\partial x^m}\right)_p\Bigg),$$

there exists in the neighborhood V a geodesic $\bar{\gamma}(s)$, where $0 \leqslant s \leqslant \sigma$, issuing from the point $\bar{p}$ and having at that point the tangent vector $\bar{B}$. If the vector $\bar{B}$ is a unit vector, this geodesic is, by definition, the segment $\gamma_{\bar{p}, \bar{B}, \sigma}$. Thus, we have proven that

Let p denote a point in V, let B denote a unit vector in M_p, and let σ denote a positive number. If the segment $\gamma_{p, B, \sigma}$ is defined and contained in V, then, for an arbitrary point $\bar{p}$ sufficiently close to the point p, an arbitrary unit vector $\bar{B} \in M_{\bar{p}}$ sufficiently close to the vector $B \in M_p$, and an arbitrary positive number $\bar{\sigma} \leqslant \sigma$, the segment $\gamma_{\bar{p}, \bar{B}, \bar{\sigma}}$ is defined and contained in V.

The terminal point $\bar{q} = \gamma_{\bar{p}, \bar{B}, \bar{\sigma}}(\bar{\sigma})$ of the segment $\gamma_{\bar{p}, \bar{B}, \bar{\sigma}}$ and the tangent vector $\bar{C} = \dot{\gamma}_{\bar{p}, \bar{B}, \bar{\sigma}}(\bar{\sigma})$ at that point have, respectively, the coordinates

$$x^i(\bar{\sigma};\ \bar{a}^1, \ldots, \bar{a}^m;\ \bar{b}^1, \ldots, \bar{b}^m)$$

and

$$\dot{x}^i(\bar{\sigma};\ \bar{a}^1, \ldots, \bar{a}^m;\ \bar{b}^1, \ldots, \bar{b}^m).$$

From this it follows immediately that

The point $\bar{q}$ and the vector $\bar{C}$ depend continuously on the point $\bar{p}$, the vector $\bar{B}$, and the number $\bar{\sigma}$.

Let us use the assertions that we have proved to prove the following

Convergence Lemma. *Let p denote an arbitrary point in a Riemannian space M and let $\{\gamma_n(t)\}$, for $0 \leqslant t \leqslant t_n$, denote a sequence of geodesics issuing from the point p on each of which the parameter t is arc length. Let us suppose that the tangent vectors $A_n = \dot{\gamma}_n(0)$ of the geodesics $\gamma_n(t)$ at the point p converge as $n \to \infty$ to some vector $A_* \in M_p$ and that the sequence $\{t_n\}$ converges to some number t_*. Then, if the maximal geodesic $\gamma_{A_*}(t)$ issuing from the point p with tangent vector A_* at the point p is defined at $t = t_*$, there obtains*

$$\gamma_{A_*}(t_*) = \lim_{n \to \infty} \gamma_n(t_n).$$

Proof: Let us partition the segment $\gamma_{A_*}|_{[0,\, t_*]}$ of the geodesic into a finite number of subsegments γ_i, for $i = 1, \ldots, r$, each of which lies in some normal convex neighborhood V_i and hence is of the form $\gamma_{p_i,\, B_i,\, \sigma_i}$. Therefore, for sufficiently large n, each segment $\gamma_n|_{[0,\, t_n]}$ is partitioned into intervals of the form $\gamma_{p_i^n,\, B_i^n,\, \sigma_i^n}$ lying in the neighborhoods V_i, where the points p_i^n are close to the points p_i, where the vectors B_i^n are close to the vectors B_i, and where $\sigma_i^n = \sigma_i$ for $i = 1, \ldots, r-1$ and $\sigma_r^n = \sigma_r + t_n - t_*$. Since, for arbitrary $i > 1$, the points p_i^n and the vectors B_i^n depend continuously on the points p_{i-1}^n, on the vectors B_{i-1}^n, and on the numbers σ_i^n, it follows that the point $\gamma_n(t_n) = p_{r+1}^n$ depends continuously on $t_n = \sum_i \sigma_i^n$. Hence, it approaches the point $\gamma_{A_*}(t_*)$. as $t_n \to t_*$.

9. Complete Riemannian spaces

We shall say that a connected Riemannian space M is complete if every maximal geodesic in it is defined on the entire real axis. It turns out that

Any two points in a complete Riemannian space can be connected by at least one minimizing curve.

Proof: Let p denote an arbitrary point in a connected complete Riemannian space M, let r denote some nonnegative number, let C_r denote the set of points $q \in M$ such that

$$\rho(p, q) \leqslant r,$$

where ρ is the interior metric on M, and let E_r denote the subset of the set C_r consisting of all points $q \in C_r$ that can be connected with the point p by at least one minimizing curve. Since, for an arbitrary point $q \in M$, there exists a number r such that $q \in C_r$, to prove the assertion made above, it suffices to show that

$$E_r = C_r \tag{1}$$

for arbitrary nonnegative r. We shall show first that

For arbitrary nonnegative r, the set E_r is compact.

Since the space M in the metric ρ is separable, to prove that the set E_r is compact, it suffices to show that from any sequence of points $q_n \in E_r$ we can choose a convergent subsequence whose limit belongs to E_r. By definition, for an arbitrary point $q_n \in E_r$, there exists a minimizing curve $\gamma_n(t)$, where $0 \leqslant t \leqslant t_n$, joining the point p with the point q_n. Assuming that the parameter t is the arc length on each of the minimizing curves γ_n, let us consider the vectors $A_n = \dot{\gamma}_n(0)$. Since the lengths of these vectors are equal to one, we may assume, turning to a subsequence of the sequence $\{q_n\}$ if necessary, that the sequence $\{A_n\}$ converges to some unit vector $A_* \in M_p$. Analogously, since $0 \leqslant t_n \leqslant r$, we may assume that the sequence $\{t_n\}$ converges to some number t_*. Here, by virtue of the assumption of completeness, the maximal geodesic $\gamma_{A_*}(t)$ is defined at $t = t_*$. Thus, the conditions of the convergence lemma proven above are satisfied for the minimizing curves γ_n (which, as we know, are geodesics). Therefore, $\gamma_n(t_n) \to \gamma_{A_*}(t_*)$; that is, $q_n \to q_*$ where $q_* = \gamma_{A_*}(t_*)$. Since $t_n = \rho(p, q_n)$, it follows that $t_* = \rho(p, q_*)$; that is, the geodesic γ_{A_*} is a minimizing curve on the segment $[0, t_*]$. Consequently, the point $q_* = \gamma_{A_*}(t_*)$ belongs to the set E_r. This completes the proof that the set E_r is compact.

Let us turn now to "induction with respect to the continuum." Obviously, equation (1) is automatically satisfied when $r = 0$. Furthermore, it is obvious that, if it is satisfied for $r = r_0$, it is also satisfied for any $r < r_0$. Finally, if it is satisfied for all $r < r_0$, it is also satisfied for $r = r_0$ since an arbitrary point $q \in C_{r_0}$ is the limit of points $q_n \in C_{r_n} = E_{r_n} \subset E_{r_0}$, where $r_n < r_0$, and the set E_{r_0}, being compact, is closed. Therefore, to prove equation (1), we need only show that if it is satisfied for some number $r \geqslant 0$, it is satisfied for some larger number $r + \delta$. Here, we may obviously assume that $C_r \neq M$.

Thus, suppose that equation (1) holds for some $r \geqslant 0$ and that $C_r \neq M$. Let us denote by $V_\delta(q)$ the spherical δ-neighborhood of the point $q \in M$ in the interior metric ρ. By virtue of the compactness of the set E_r, there exists a sequence of points $q_1, \ldots, q_n$ and positive numbers $\delta_1, \ldots, \delta_n$ such that:

(1) the neighborhoods $V_{\delta i}(q_i)$ cover the set E_r;

(2) the closure of each neighborhood $V_{2\delta_i}(q_i)$ is compact;

(3) each neighborhood $V_{2\delta_i}(q_i)$ is a normal convex neighborhood of the point q_i.

Define

$$V = V_{\delta_1}(q_1) \cup \ldots \cup V_{\delta_n}(q_n).$$

and let $\dot{V}$ denote the boundary of the open set V. Since the sets E_r and $\dot{V}$ are compact and disjoint (because $E_r \subset V$), the function $\rho(q, \bar{q})$, where $q \in E_r$ and $\bar{q} \in \dot{V}$, has a positive greatest lower bound ρ_0. An arbitrary piecewise-regular curve connecting the point p with an arbitrary point $q \notin V$ crosses the boundary S_r of the set $E_r = C_r$ and the boundary $\dot{V}$ of the set V (or it stops at $\dot{V}$). Therefore, its length is not less than $r + \rho_0$. This means that

$$\rho(p, q) \geqslant r + \rho_0.$$

Therefore, for arbitrary positive $\delta < \rho_0$, the set $C_{r+\delta}$ is entirely contained in the set V. Let us show that for $\delta < \min(\delta_1, \ldots, \delta_n)$, we have

$$E_{r+\delta} = C_{r+\delta}. \tag{2}$$

Suppose that $q \in C_{r+\delta}$. If $q \in C_r$, then $q \in E_r \subset E_{r+\delta}$. Therefore, we may assume that $q \notin C_r$. Let q' denote a point in the set S_r that lies at minimal distance from the point q (the existence of such a point being ensured by the compactness of the set $S_r \subset E_r$). Since an arbitrary piecewise-regular curve connecting the points p and q intersects the boundary S_r, its length is not less than $r + \rho(q, q')$. This means that

$$\rho(p, q) \geqslant r + \rho(q, q')$$

and, hence,

$$\rho(q, q') \leqslant \rho(p, q) - r \leqslant \delta.$$

Since $\delta < \min(\delta_1, \ldots, \delta_n)$, it follows that the points q and q' lie in the same normal convex neighborhood $V_{2\delta i}(q_i)$. This is true because the fact that the neighborhoods $V_{\delta i}(q_i)$ cover E_r implies the existence of an i such that $q \in V_{\delta_i}(q_i)$ and hence

$$q' \in V_\delta(q) \subset V_{2\delta_i}(q_i).$$

Consequently, the points q and q' can be connected by a minimizing curve. If we combine this minimizing curve with the minimizing curve connecting the points p and q', we obtain a curve of length

$$r + \rho(q, q') \leqslant \rho(p, q),$$

that is, of length $\rho(p, q)$, and hence a minimizing curve connecting the points p and q. Consequently, $q \in E_{r+\delta}$. This completes the proof of equation (2) and hence of the assertion made above.

Remark: In proving the compactness of the set E_r, we used only its boundedness (in the metric ρ), the fact that $q_* = \gamma_{A*}(t_*) \in E_r$, and, of course, the assumption that an arbitrary point in it can be connected to the point p by at least one minimizing curve. On the basis of the assertion proven, this last condition is satisfied for an arbitrary subset E of the entire space M. Since the equation $q_* = \lim_{n\to\infty} q_n$, the relation $q_* \in E$, and the fact that the set E is closed together imply that $q_n \in E$, we conclude that

An arbitrary bounded closed subset I of a complete Riemannian space M is compact.

For an arbitrary connected Riemannian space M, we can regard the distance $\rho(p, q)$ as a function defined on the Cartesian product $M \times M$ (cf. section 15 of Chapter 1) of the manifold M with itself.

It turns out that

If the points p and q are distinct, the distance ρ is a smooth function at the point $(p, q) \in M \times M$ if either

(1) the points p and q are sufficiently close (contained in some single normal convex neighborhood)

or

(2) the space M is complete.

Proof: Suppose that condition (1) is satisfied, that is, that the points p and q are contained in some normal convex neighborhood V. Then, the distance $\rho(p, q)$ is equal to the length of the geodesic connecting in V the point p with the point q. But we know that this length is equal to the length $|A|$ of the vector $A \in M_p$ for which Exp $A = q$. Since $q \neq p$ and hence $A \neq 0$, the length $|A|$ is a smooth function of the coordinates of the vector A, that is, of the normal coordinates of the point q. Thus, the distance $\rho(p, q)$ depends smoothly on the point q. Analogous reasoning shows that it depends smoothly on the point p also. Consequently, the function ρ is smooth at the point $(p, q) \in M \times M$.

Suppose now that condition (2) is satisfied, that is, that M is complete. Then, from what we have proven above, there exists in M a minimizing geodesic $\gamma(t)$, for $\alpha \leqslant t \leqslant \beta$, connecting the point p with the point q. Since the segment $[\alpha, \beta]$ is compact, there exist numbers

$$\alpha = t_0 < t_1 < \ldots < t_n < t_{n+1} = \beta,$$

such that, for arbitrary $i = 0, \ldots, n$, the points $p_i = \gamma(t_i)$ and $p_{i+1} = \gamma(t_{i+1})$ [obviously distinct] belong to a single normal convex neighborhood. Also, from what we have just proved, the distance $\rho(p_i, p_{i+1})$ will be a smooth function of the points p_i and p_{i+1}. But then, the number

$$\rho(p, q) = \sum_{i=0}^{n} \rho(p_i, p_{i+1})$$

will be a smooth function of the points $p_0, \ldots, p_{n+1}$ and, in particular, of the points $p = p_0$ and $q = p_{n+1}$.

At points of the form (p, p), the function $\rho(p, q)$ is not smooth. However, it is obvious that its square $\rho^2(p, q)$ is smooth at these points (since the square $|A|^2$ of the length $|A|$ of the vector A is smooth even at $A = 0$). Since, in accordance with what we have proven, it is also smooth at all other points of the Cartesian product $M \times M$ (we are assuming that the space M is complete), it follows that

For an arbitrary complete Riemannian space M, the function $\rho^2(p, q)$ *is a smooth function on the manifold* $M \times M$.

10. Conditions for completeness of Riemannian spaces

Quite unexpected is the fact that complete Riemannian spaces can be characterized in the class of all Riemannian spaces by extremely simple essentially topological properties. Specifically,

For an arbitrary connected Riemannian space M, the following three properties are equivalent:

(1) the space M is complete in the sense of section 9; that is, each maximal geodesic in it is defined on the entire real axis;

(2) the space M is complete with respect to the interior metric ρ*; that is, every fundamental sequence of points in it converges to a point in it;*

(3) every bounded (in the metric ρ*) closed subset of the space M is compact.*

That property (1) implies property (3) was noted in section 9; that property (3) implies property (2) is trivial and self-evident. This is true because the closure of an arbitrary fundamental sequence of points p_n in the space M is obviously bounded and, hence, by virtue of property (3), compact. Therefore, the sequence $\{p_n\}$ has a cluster point, which, by virtue of the assumption that the sequence is fundamental, is its limit.

Thus, it only remains to prove that property (2) [i.e., metric completeness] implies property (1) [i.e., completeness in the sense of section 9].

Let $\gamma : I \to M$ denote an arbitrary geodesic of a metrically complete Riemannian space. Suppose that γ is defined on an interval $I = (a, b)$ that is finite on the right. For an arbitrary sequence $\{t_n\}$ of numbers $t_n \in I$ that converges to the number b, the sequence $\{p_n\}$, where $p_n = \gamma(t_n)$ is a fundamental sequence in M (by virtue of our assumption that the parameter t is arc length $\rho(p_i, p_j) = |t_i - t_j|$) and hence has a definite limit p in M. Obviously, this limit is independent of the choice of the sequence $\{t_n\}$. Let V denote an arbitrary normal convex neighborhood of the point p and let $x^1, \ldots, x^m$ denote a system of local coordinates defined in the neighborhood V. Consider the functions $x^i(t) = x^i(\gamma(t))$ that define the geodesic $\gamma(t)$ parametrically in the neighborhood V. These functions are defined on some interval $I_0 \subset I$ with right end-point b. Here, the limit $\lim\limits_{t \to b-0} x^i(t)$ exists for arbitrary i. Let us extend the definition of the function $x^i(t)$ to the point b by defining

$$x^i(b) = \lim_{t \to b-0} x^i(t).$$

Since the functions $x^i(t)$ satisfy on the interval I_0 the differential equations of the geodesics

$$\ddot{x}^i + \Gamma^i_{jk}(x^1, \ldots, x^m)\,\dot{x}^j \dot{x}^k = 0, \tag{1}$$

and since the first derivatives $\dot{x}^i(t)$ are bounded on I_0 (because

$$\sum_i [\dot{x}^i(t)]^2 = 1$$

for arbitrary $t \in I_0$), the second derivatives $\ddot{x}^i(t)$ are also bounded on I_0. Therefore, the first derivatives $\dot{x}^i(t)$ are uniformly continuous on I_0. It follows from this that, in particular, they have a definite limit $\lim_{t \to b-0} \dot{x}^i(t)$ as $t \to b-0$. Since, in accordance with Lagrange's theorem,

$$\frac{x^i(t) - x^i(b)}{t-b} = \dot{x}^i(\xi), \text{ where } t \leqslant \xi \leqslant b,$$

this limit coincides with the left-hand derivative $\dot{x}^i(b)$ of the function $x^i(t)$ at the point b:

$$\dot{x}^i(b) = \lim_{t \to b-0} \dot{x}^i(t). \tag{2}$$

By virtue of equations (1), existence of the limit $\lim_{t \to b-0} \dot{x}^i(t)$ implies existence of the limit $\lim_{t \to b-0} \ddot{x}^i(t)$. If we apply Lagrange's theorem to the function $\dot{x}^i(t)$, we see, just as above, that this limit coincides with the left-hand derivative $\ddot{x}^i(b)$ of the function $\dot{x}^i(t)$ at the point b:

$$\ddot{x}^i(b) = \lim_{t \to b-0} \ddot{x}^i(b). \tag{3}$$

Let us now set

$$B = \dot{x}^i(b)\left(\frac{\partial}{\partial x^i}\right)_p,$$

and consider the geodesic γ_B issuing from the point B and tangent at that point to the vector p. Suppose that the parameter t ranges over the half-closed, half-open interval $b \leqslant t < c$ on this geodesic. Let us define on M a continuous curve $\bar{\gamma}: (a, c) \to M$ by

$$\bar{\gamma}(t) = \begin{cases} \gamma(t), & \text{if} \quad a < t < b, \\ \gamma_B(t), & \text{if} \quad b \leqslant t < c. \end{cases}$$

For $t \neq b$, the functions $\bar{x}^i(t)$, which define this curve parametrically in the neighborhood V, satisfy equations (1). Furthermore, at the point b, the derivatives $\dot{\bar{x}}^i(b) = \dot{x}^i(b)$ exist. Also, since the curve $\gamma_B(t)$ is a geodesic, the right-hand derivatives $\ddot{\bar{x}}(b)$ exist and satisfy equations (1). But, by virtue of equations (2) and (3), the left-hand derivatives $\ddot{\bar{x}}^i(b) = \ddot{x}^i(b)$ also satisfy these equations. Therefore, the functions $x^i(t)$ have ordinary second derivatives at the point b, and these derivatives are continuous and satisfy the system (1). As we know (cf. section 6 of Chapter 2), it follows from this that the curve $\bar{\gamma}$ is regular and is a geodesic.

Thus, an arbitrary geodesic γ defined on an interval (a, b) that is finite on the right is the restriction of some geodesic $\bar{\gamma}$ defined on a larger interval (a, c). Analogously, one can show that if the interval (a, b) is finite on the left, the geodesic γ can be extended to a larger interval (c, b), where $c < a$. Consequently, every maximal geodesic is necessarily defined on the entire real axis. This completes the proof of the equivalence of properties (1), (2) and (3).

In particular, it follows from the assertion just proven that

An arbitrary compact connected Riemannian space is complete.

CHAPTER 4

THE VARIATIONAL PROPERTIES OF GEODESICS

1. Geodesics as curves of stationary length

In general, a geodesic is not a minimizing curve; that is, its length between two arbitrary points on it is not minimum. We can only assert that the length attains its extreme (stationary) value on a geodesic. The present section is devoted primarily to proof of this assertion, the precise formulation of which is given below.

Let $\gamma : [t_1, t_2] \to M$ denote an arbitrary regular curve in a Riemannian space M (more precisely, a segment of such a curve) and let p and q denote its end-points:

$$p = \gamma(t_1), \quad q = \gamma(t_2).$$

Let G denote the subset of the plane R^2 consisting of all points $(t, \tau) \in R^2$ at which

$$t_1(\tau) \leqslant t \leqslant t_2(\tau),$$
$$-\tau_0 < \tau < \tau_0,$$

where τ_0 is some positive number and $t_1(\tau)$ and $t_2(\tau)$ are smooth functions defined for $|\tau| < \tau_0$ such that $t_1(\tau) < t_2(\tau)$, $t_1(0) = t_1$, and $t_2(0) = t_2$.

We shall call the surface

$$\varphi : G \to M$$

(more precisely, the restriction to G of some surface defined on a region in R^2 that contains the set G) a *variation* of the curve γ if the curve

$$\varphi_0(t) = \varphi(t, 0)$$

coincides with the curve $\gamma(t)$. Every variation φ defines on M two (in general, nonregular) bounding curves

$$\varphi_p(\tau) = \varphi(t_1(\tau), \tau), \quad \varphi_q(\tau) = \varphi(t_2(\tau), \tau),$$

passing through the points p and q, respectively:

$$\varphi_p(0) = p, \quad \varphi_q(0) = q.$$

If these curves are degenerate, that is, if

$$\varphi_p(\tau) = p, \quad \varphi_q(\tau) = q$$

for all τ, we shall call the surface φ a variation with fixed end-points.

At every point $\varphi(t, \tau) \in M$ at which $(t, \tau) \in G$, the variation φ defines two vectors $\frac{\partial \varphi}{\partial t}(t, \tau)$ and $\frac{\partial \varphi}{\partial \tau}(t, \tau)$. In particular, at every point $\gamma(t)$ of the curve γ, the variation φ defines the vectors $\frac{\partial \varphi}{\partial t}(t, 0)$ and $\frac{\partial \varphi}{\partial \tau}(t, 0)$. Here, we obviously have

$$\frac{\partial \varphi}{\partial t}(t, 0) = \dot{\gamma}(t).$$

The vectors $\frac{\partial \varphi}{\partial \tau}(t, 0)$ constitute a new vector field

$$X(t) = \frac{\partial \varphi}{\partial \tau}(t, 0)$$

on the curve $\gamma(t)$. We shall say that this vector field is *associated* with the variation φ. It is connected with the tangent vectors $\dot{\varphi}_p(0)$ and $\dot{\varphi}_q(0)$ to the curves $\varphi_p(\tau)$ and $\varphi_q(\tau)$ at the points p and q, respectively, by the formulas

$$\dot{\varphi}_p(0) = \dot{t}_1(0)\,\dot{\gamma}(t_1) + X(t_1),$$
$$\dot{\varphi}_q(0) = \dot{t}_2(0)\,\dot{\gamma}(t_2) + X(t_2).$$

Consequently,

If φ is a variation with fixed end points, then

$$X(t_1) = -\dot{t}_1(0)\,\dot{\gamma}(t_1),$$
$$X(t_2) = -\dot{t}_2(0)\,\dot{\gamma}(t_2).$$

In particular,

If $t_1(\tau) = t_1$ and $t_2(\tau) = t_2$ for all τ (that is, if the set G is a rectangle), then, for an arbitrary variation with fixed end-points,

$$X(t_1) = X(t_2) = 0.$$

Obviously, this necessary condition is also sufficient; that is,

For an arbitrary vector field $X(t)$ on the curve $\gamma(t)$ such that

$$X(t_1) = X(t_2) = 0,$$

there exists a variation φ with fixed end-points that is defined on *the rectangle*

$$t_1 \leqslant t \leqslant t_2, \quad -\tau_0 < \tau < \tau_0,$$

where τ_0 *is some (sufficiently small) positive number, and to which the field* $X(t)$ *is associated.*

Every variation φ defines, for each τ, the curve

$$\varphi_\tau(t) = \varphi(t, \tau), \quad t_1(\tau) \leqslant t \leqslant t_2(\tau)$$

with tangent vector

$$\dot{\varphi}_\tau(t) = \frac{\partial \varphi}{\partial t}(t, \tau).$$

Suppose that

$$J(\tau) = \int_{t_1(\tau)}^{t_2(\tau)} |\dot{\varphi}_\tau(t)|\, dt$$

is the length of this curve. If we differentiate the function $J(\tau)$ with respect to τ and note that, in accordance with formula (8) of section 1 of Chapter 3, the derivative with respect to τ of the function

$$|\dot{\varphi}_\tau(t)| = \sqrt{\left(\frac{\partial \varphi}{\partial t}(t, \tau), \frac{\partial \varphi}{\partial t}(t, \tau)\right)}$$

is expressed by the formula

$$\frac{d}{dt}|\dot{\varphi}_\tau(t)| = \frac{1}{\left|\frac{\partial \varphi}{\partial t}(t, \tau)\right|}\left(\frac{\nabla}{\partial \tau}\frac{\partial \varphi}{\partial t}(t, \tau), \frac{\partial \varphi}{\partial t}(t, \tau)\right), \tag{1}$$

then, by using formula (2) of section 3 of Chapter 2, we see that

$$\dot{J}(\tau) = \dot{t}_2(\tau)\left|\frac{\partial \varphi}{\partial t}(t_2(\tau), \tau)\right| - \dot{t}_1(\tau)\left|\frac{\partial \varphi}{\partial t}(t_1(\tau), \tau)\right| + \\ + \int_{t_1(\tau)}^{t_2(\tau)} \frac{1}{\left|\frac{\partial \varphi}{\partial t}(t, \tau)\right|}\left(\frac{\nabla}{\partial t}\frac{\partial \varphi}{\partial \tau}(t, \tau), \frac{\partial \varphi}{\partial t}(t, \tau)\right) dt.$$

If we set $\tau = 0$ in this equation and assume, for simplicity, that $|\dot{\gamma}(t)| = 1$, that is, that the parameter t is arc length on the curve $\gamma(t)$, we obtain the formula

$$\dot{J}(0) = \dot{t}_2(0) - \dot{t}_1(0) + \int_{t_1}^{t_2}\left(\frac{\nabla X}{dt}(t), \dot{\gamma}(t)\right) dt.$$

But, in accordance with formula (8) of section 1 of Chapter 3,

$$\left(\frac{\nabla X}{dt}(t), \dot{\gamma}(t)\right) = \frac{d}{dt}(X(t), \dot{\gamma}(t)) - \left(X(t), \frac{\nabla \dot{\gamma}}{dt}(t)\right),$$

so that

$$\dot{j}(0) = \dot{t}_2(0) - \dot{t}_1(0) + (X(t_2), \dot{\gamma}(t_2)) - (X(t_1), \dot{\gamma}(t_1)) - \\ - \int_{t_1}^{t_2} \left(X(t), \frac{\nabla\dot{\gamma}}{dt}(t)\right) dt.$$

On the basis of what we have proved above, it follows immediately from this that

$$\dot{j}(0) = (\dot{\varphi}_q(0), \dot{\gamma}_q) - (\dot{\varphi}_p(0), \dot{\gamma}_p) - \int_{t_1}^{t_2} \left(X(t), \frac{\nabla\dot{\gamma}}{dt}(t)\right) dt, \tag{2}$$

where $\dot{\gamma}_q = \dot{\gamma}(t_2)$ and $\dot{\gamma}_p = \dot{\gamma}(t_1)$ are the tangent vectors to the curve γ at the points q and p, respectively. In particular,

For an arbitrary variation φ with fixed end-points,

$$\dot{j}(0) = - \int_{t_1}^{t_2} \left(X(t), \frac{\nabla\dot{\gamma}}{dt}(t)\right) dt. \tag{3}$$

The number $\dot{j}(0)$ is called the *first variation* of the arc length of the curve γ corresponding to the given variation φ of the curve γ. If it is equal to 0 for every variation φ with fixed end-points, then the curve γ is called a *curve of stationary length.* It follows immediately from formula (3) that

An arbitrary geodesic $\gamma(t)$ is a curve of stationary length.

Remark: The converse is also true; that is,

An arbitrary curve of stationary length and for which the parameter is arc length is a geodesic.

Proof: On the basis of the so-called fundamental lemma of the calculus of variations (for the application of which, the scalar product

$$\left(X(t), \frac{\nabla\dot{\gamma}}{dt}(t)\right)$$

at every point $t \in [t_1, t_2]$ should be expressed in terms of local coordinates), it follows from the condition $\dot{j}(0) = 0$ that $\frac{\nabla\dot{\gamma}}{dt}(t) = 0$, (because, for $t_1 < t < t_2$, the vector $X(t)$ can, as we have noted, be chosen arbitrarily); that is, the curve $\gamma(t)$ is a geodesic.

For an arbitrary variation φ of a geodesic γ, it follows from formula (2) that

$$\dot{j}(0) = (\dot{\varphi}_q(0), \dot{\gamma}_q) - (\dot{\varphi}_p(0), \dot{\gamma}_p). \tag{4}$$

We shall have more than one occasion in what follows to use this important formula.

Let us consider, for example, the function $\rho=\rho(p, q)$ on the manifold $M\times M$. As we know (cf. section 9 of Chapter 3), the function ρ is smooth at the point $(p, q)\in M\times M$ if the points p and q are distinct though sufficiently close to one another (that is, contained in a single normal convex neighborhood V). Therefore, for an arbitrary vector $C\in(M\times M)_{(p, q)}$, the number $C\rho$ is defined.

Define $C=(C_1, C_2)$, where $C_1\in M_p$ and $C_2\in M_q$ (cf. section 15 of Chapter 1). Consider an arbitrary curve $\varphi(\tau)$, where $|\tau|<\tau_0$, in the manifold $M\times M$ for which

$$\varphi(0)=(p, q), \quad \dot{\varphi}(0)=C.$$

Suppose that $\varphi(\tau)=(\varphi_1(\tau), \varphi_2(\tau))$, where $\varphi_1(\tau), \varphi_2(\tau)\in M$. Then,

$$\varphi_1(0)=p, \quad \dot{\varphi}_1(0)=C_1,$$
$$\varphi_2(0)=q, \quad \dot{\varphi}_2(0)=C_2.$$

Let us define a real function $\rho(\tau)$ by

$$\rho(\tau)=\rho(\varphi(\tau))=\rho(\varphi_1(\tau), \varphi_2(\tau)).$$

In accordance with the remark at the end of section 5 of Chapter 2, the number $C\rho$ is equal to the value $\dot{\rho}(0)$ of the derivative $\dot{\rho}(\tau)$ of this function at $\tau=0$:

$$C\rho=\dot{\rho}(0).$$

To evaluate this derivative, we note that, for sufficiently small positive τ_0, all points on the curves $\varphi_1(\tau)$ and $\varphi_2(\tau)$ belong to a neighborhood V containing the points p and q. Therefore, for arbitrary $\tau\in(-\tau_0, \tau_0)$ the points $\varphi_1(\tau)$ and $\varphi_2(\tau)$ can be connected in M by a unique minimizing curve $\gamma_\tau(t)$. Taking arc length on this minimizing curve [measured from the point $\varphi_1(t)$] as the parameter t, we see immediately that the formula

$$\varphi(t, \tau)=\gamma_\tau(t), \quad |\tau|<\tau_0, \quad 0\leqslant t\leqslant\rho(\tau)$$

defines a variation $\varphi(t, \tau)$ of the minimizing geodesic $\gamma=\gamma_0$ connecting the points p and q. Obviously, the curves $\varphi_1(\tau)$ and $\varphi_2(\tau)$ will serve as the bounding curves $\varphi_p(\tau)$ and $\varphi_q(\tau)$ of this variation, and the function $J(\tau)$ corresponding to it coincides with the function $\rho(\tau)$. Therefore, in accordance with formula (4),

$$\dot{\rho}(0)=(\dot{\varphi}_2(0), \dot{\gamma}_q)-(\dot{\varphi}_1(0), \dot{\gamma}_p)=(C_2, \dot{\gamma}_q)-(C_1, \dot{\gamma}_p).$$

Thus, we have proven that

If two distinct points p and q of a Riemannian space M are sufficiently close to each other, then, for an arbitrary vector

$$C=(C_1, C_2)\in(M\times M)_{(p, q)},$$

where $C_1 \in M_p$ and $C_2 \in M_q$, we have

$$C\rho = (C_2, \dot{\gamma}_q) - (C_1, \dot{\gamma}_p), \tag{5}$$

where $\dot{\gamma}_p$ and $\dot{\gamma}_q$ are unit tangent vectors at points p and q of the minimizing geodesic γ connecting these two points.

We note in conclusion that we may use formula (4) without the assumption that the parameter t on the geodesic γ is arc length. Obviously, in this case, formula (4) takes the form

$$\dot{J}(0) = \frac{(\dot{\varphi}_q(0), \dot{\gamma}_q) - (\dot{\varphi}_p(0), \dot{\gamma}_p)}{l}, \tag{6}$$

where $l = |\dot{\gamma}(t)|$ (as we know, this number is independent of t) and $\dot{\gamma}_p$ and $\dot{\gamma}_q$ are, as above, the tangent vectors $\dot{\gamma}(t_1)$ and $\dot{\gamma}(t_2)$ to the geodesic γ at the points p and q, respectively.

2. The second variation of arc length of a geodesic

In what follows, we shall also need the second variation $\ddot{J}(0)$ of the arc length of the curve $\gamma(t)$, though only for the case in which the curve $\gamma(t)$ is a geodesic. Let us evaluate this variation under the additional assumption (which is quite sufficient for our purposes) that

(α) *For an arbitrary point $(t, \tau) \in G$, the length*

$$k(\tau) = \left|\frac{\partial \varphi}{\partial t}(t, \tau)\right|$$

of the vector $\frac{\partial \varphi}{\partial t}(t, \tau) = \dot{\varphi}_\tau(t)$ is nonzero and depends only on τ, not on t.

This assumption means that every curve $\varphi_\tau(t)$ is regular and that the parameter t on it is proportional to arc length with a positive coefficient of proportionality $k(\tau)$ depending only on τ.

By virtue of this assumption,

$$J(\tau) = k(\tau)(t_2(\tau) - t_1(\tau)).$$

Therefore,

$$\ddot{J}(\tau) = \ddot{k}(\tau)(t_2(\tau) - t_1(\tau)) + 2\dot{k}(\tau)(\dot{t}_2(\tau) - \dot{t}_1(\tau)) + \\ + k(\tau)(\ddot{t}_2(\tau) - \ddot{t}_1(\tau)),$$

so that

$$\ddot{J}(0) = \ddot{k}(0)(t_2(0) - t_1(0)) + 2\dot{k}(0)(\dot{t}_2(0) - \dot{t}_1(0)) + \\ + k(0)(\ddot{t}_2(0) - \ddot{t}_1(0)).$$

But as we have seen (cf. formula (1) of section 1),

$$\dot{k}(\tau) = \frac{1}{k(\tau)}\left(\frac{\nabla}{\partial \tau}\frac{\partial \varphi}{\partial t}(t, \tau), \frac{\partial \varphi}{\partial t}(t, \tau)\right),$$

from which it follows that

$$\ddot{k}(\tau)=\frac{1}{k(\tau)}\Big[\Big(\frac{\nabla}{\partial\tau}\frac{\nabla}{\partial\tau}\frac{\partial\varphi}{\partial t}(t,\tau),\ \frac{\partial\varphi}{\partial t}(t,\tau)\Big)+ +\Big(\frac{\nabla}{\partial\tau}\frac{\partial\varphi}{\partial t}(t,\tau),\ \frac{\nabla}{\partial\tau}\frac{\partial\varphi}{\partial t}(t,\tau)\Big)\Big]- -\frac{1}{k(\tau)^3}\Big(\frac{\nabla}{\partial\tau}\frac{\partial\varphi}{\partial t}(t,\tau),\ \frac{\partial\varphi}{\partial t}(t,\tau)\Big)^2.$$

Thus,

$$\dot{k}(0)=\frac{1}{k(0)}\Big(\frac{\nabla}{\partial\tau}\frac{\partial\varphi}{\partial t}(t,0),\ \frac{\partial\varphi}{\partial t}(t,0)\Big),$$

$$\ddot{k}(0)=\frac{1}{k(0)}\Big[\Big(\frac{\nabla}{\partial\tau}\frac{\nabla}{\partial\tau}\frac{\partial\varphi}{\partial t}(t,0),\ \frac{\partial\varphi}{\partial t}(t,0)\Big)+ +\Big(\frac{\nabla}{\partial\tau}\frac{\partial\varphi}{\partial t}(t,0),\ \frac{\nabla}{\partial\tau}\frac{\partial\varphi}{\partial t}(t,0)\Big)\Big]- -\frac{1}{k(0)^3}\Big(\frac{\nabla}{\partial\tau}\frac{\partial\varphi}{\partial t}(t,0),\ \frac{\partial\varphi}{\partial t}(t,0)\Big)^2.$$

But, in accordance with formulas (2) and (3) of section 3 of Chapter 2,

$$\frac{\nabla}{\partial\tau}\frac{\partial\varphi}{\partial t}(t,\tau)=\frac{\nabla}{\partial t}\frac{\partial\varphi}{\partial\tau}(t,\tau),$$

$$\frac{\nabla}{\partial\tau}\frac{\nabla}{\partial\tau}\frac{\partial\varphi}{\partial t}(t,\tau)=\frac{\nabla}{\partial t}\frac{\nabla}{\partial\tau}\frac{\partial\varphi}{\partial\tau}(t,\tau)+ +R_{\varphi(t,\tau)}\Big(\frac{\partial\varphi}{\partial\tau}(t,\tau),\ \frac{\partial\varphi}{\partial t}(t,\tau)\Big)\frac{\partial\varphi}{\partial\tau}(t,\tau)$$

and, hence,

$$\frac{\nabla}{\partial\tau}\frac{\partial\varphi}{\partial t}(t,0)=\frac{\nabla X}{dt}(t),$$

$$\frac{\nabla}{\partial\tau}\frac{\nabla}{\partial\tau}\frac{\partial\varphi}{\partial t}(t,0)=\frac{\nabla}{\partial t}\frac{\nabla}{\partial\tau}\frac{\partial\varphi}{\partial\tau}(t,0)+R_{\gamma(t)}(X(t),\ \dot{\gamma}(t))X(t).$$

Thus,

$$\dot{k}(0)=\frac{1}{k(0)}\Big(\frac{\nabla X}{dt}(t),\ \dot{\gamma}(t)\Big) \tag{1}$$

(we have already derived what is essentially this formula) and

$$\ddot{k}(0)=\frac{1}{k(0)}\Big[\Big(\frac{\nabla}{\partial t}\frac{\nabla}{\partial\tau}\frac{\partial\varphi}{\partial\tau}(t,0),\ \dot{\gamma}(t)\Big)+ +R_{\gamma(t)}(X(t),\ \dot{\gamma}(t),\ X(t),\ \dot{\gamma}(t))+\Big(\frac{\nabla X}{dt}(t),\ \frac{\nabla X}{dt}(t)\Big)\Big]- -\frac{1}{k(0)^3}\Big(\frac{\nabla X}{dt}(t),\ \dot{\gamma}(t)\Big)^2.$$

Since $\ddot{k}(0)$ is independent of t and since $t_2(0)=t_2$ and $t_1(0)=t_1$, we have

$$\ddot{k}(0)(t_2(0)-t_1(0))=\int_{t_1}^{t_2}\ddot{k}(0)\,dt.$$

On the other hand, since the curve $\gamma(t)$ is a geodesic (this is the first time we use this assumption), we have

$$\frac{\nabla}{dt}\dot{\gamma}(t)=0$$

and, therefore

$$\left(\frac{\nabla}{\partial t}\frac{\nabla}{\partial \tau}\frac{\partial \varphi}{\partial \tau}(t,0),\ \dot{\gamma}(t)\right)=\frac{d}{dt}\left(\frac{\nabla}{\partial \tau}\frac{\partial \varphi}{\partial \tau}(t,0),\ \dot{\gamma}(t)\right).$$

Consequently, if we permute the arguments of the Riemannian tensor, we obtain

$$\begin{aligned}\ddot{k}(0)(t_2(0)-t_1(0))&=\\ &=\frac{1}{k(0)}\left[\left(\frac{\nabla}{\partial \tau}\frac{\partial \varphi}{\partial \tau}(t_2,0),\ \dot{\gamma}(t_2)\right)-\left(\frac{\nabla}{\partial \tau}\frac{\partial \varphi}{\partial \tau}(t_1,0),\ \dot{\gamma}(t_1)\right)\right]+\\ &\quad+\int_{t_1}^{t_2}\left\{\frac{1}{k(0)}\left[\left(\frac{\nabla X}{dt}(t),\ \frac{\nabla X}{dt}(t)\right)-\right.\right.\\ &\left.\left.-R_{\gamma(t)}(X(t),\ \dot{\gamma}(t),\ \dot{\gamma}(t),\ X(t))\right]-\frac{1}{k^3(0)}\left(\frac{\nabla X}{dt}(t),\ \dot{\gamma}(t)\right)^2\right\}dt.\end{aligned}$$

Let us now consider the bounding curves $\varphi_p(\tau)$ and $\varphi_q(\tau)$ of the variation φ. Since

$$\dot{\varphi}_p(\tau)=\frac{\partial \varphi}{\partial t}(t_1(\tau),\ \tau)\,\dot{t}_1(\tau)+\frac{\partial \varphi}{\partial \tau}(t_1(\tau),\ \tau),$$

we have

$$\begin{aligned}\frac{\nabla\dot{\varphi}_p}{d\tau}(\tau)=\frac{\nabla}{\partial t}\frac{\partial \varphi}{\partial t}(t_1(\tau),\ \tau)\,[\dot{t}_1(\tau)]^2+2\frac{\nabla}{\partial \tau}\frac{\partial \varphi}{\partial t}(t_1(\tau),\ \tau)\,\dot{t}_1(\tau)+\\ +\frac{\partial \varphi}{\partial t}(t_1(\tau),\ \tau)\,\ddot{t}_1(\tau)+\frac{\nabla}{\partial \tau}\frac{\partial \varphi}{\partial \tau}(t_1(\tau),\ \tau),\end{aligned}$$

and hence

$$\frac{\nabla\dot{\varphi}_p}{d\tau}(0)=2\frac{\nabla X}{dt}(t_1)\,\dot{t}_1(0)+\dot{\gamma}(t_1)\,\ddot{t}_1(0)+\frac{\nabla}{\partial \tau}\frac{\partial \varphi}{\partial \tau}(t_1,\ 0),$$

since $\frac{\nabla\dot{\gamma}}{dt}(t)=0$. Consequently,

$$\begin{aligned}\left(\frac{\nabla}{\partial \tau}\frac{\partial \varphi}{\partial \tau}(t_1,\ 0),\ \dot{\gamma}(t_1)\right)&=\left(\frac{\nabla\dot{\varphi}_p}{d\tau}(0),\ \dot{\gamma}(t_1)\right)-\\ &-2\left(\frac{\nabla X}{dt}(t_1),\ \dot{\gamma}(t_1)\right)\dot{t}_1(0)-(\dot{\gamma}(t_1),\ \dot{\gamma}(t_1))\,\ddot{t}_1(0)=\\ &=\left(\frac{\nabla\dot{\varphi}_p}{d\tau}(0),\ \dot{\gamma}(t_1)\right)-2\dot{k}(0)\,k(0)\,\dot{t}_1(0)-k(0)^2\,\ddot{t}_1(0).\end{aligned}$$

(In this last transformation, we used the fact that the numbers $k(0)$

and $\dot{k}(0)$ are independent of t, so that

$$k(0)=|\dot{\gamma}(t_1)| \quad \text{and} \quad \dot{k}(0)=\frac{1}{k(0)}\left(\frac{\nabla X}{dt}(t_1),\ \dot{\gamma}(t_1)\right).$$

Obviously, an analogous formula holds for

$$\left(\frac{\nabla}{\partial\tau}\frac{\partial\varphi}{\partial\tau}(t_2,\ 0),\ \dot{\gamma}(t_2)\right).$$

If we substitute these expressions into the formula for

$$\ddot{k}(0)(t_2(0)-t_1(0)),$$

we immediately obtain

The second variation $\ddot{J}(0)$ *of the arc length of a geodesic* $\gamma(t)$ *for an arbitrary variation* φ *that satisfies hypothesis* (α) *is expressed by the formula*

$$\begin{aligned}\ddot{J}(0)=\frac{1}{k(0)}\left[\left(\frac{\nabla\dot{\varphi}_q}{d\tau}(0),\ \dot{\gamma}_q\right)-\left(\frac{\nabla\dot{\varphi}_p}{d\tau}(0),\ \dot{\gamma}_p\right)\right]+\\ +\int_{t_1}^{t_2}\left\{\frac{1}{k(0)}\left[\left(\frac{\nabla X}{dt}(t),\ \frac{\nabla X}{dt}(t)\right)-\right.\right.\\ \left.\left.-R_{\gamma(t)}(X(t),\ \dot{\gamma}(t),\ \dot{\gamma}(t),\ X(t))\right]-\frac{1}{k(0)^3}\left(\frac{\nabla X}{dt}(t),\ \dot{\gamma}(t)\right)^2\right\}dt.\end{aligned} \tag{2}$$

3. Jacobi variations and Jacobi fields

Now, let M denote a *complete* Riemannian space and let $\gamma: R\to M$ denote an arbitrary maximal nondegenerate geodesic in it. We shall refer to the variation

$$\varphi: R\times I\to M$$

of the geodesic γ, where this variation is defined in the strip $R\times I\subset R^2$, where I in turn is an interval of the form $(-\tau_0,\ \tau_0)$, as a *Jacobi variation* if, for arbitrary $\tau\in I$, the curve

$$\varphi_\tau(t)=\varphi(t,\ \tau), \qquad t\in R$$

is a geodesic.

For every fixed $t_0\in R$, the Jacobi variation φ defines a smooth (though not, in general, regular) curve

$$\alpha(\tau)=\varphi(t_0,\ \tau), \qquad \tau\in I$$

and a vector field

$$A(\tau)=\frac{\partial\varphi}{\partial t}(t_0,\ \tau), \qquad \tau\in I$$

on the curve $\alpha(\tau)$. Since

$$A(\tau) = \dot{\varphi}_\tau(t_0),$$

the vector $A(\tau)$ unambiguously determines the geodesic $\varphi_\tau(t)$ for arbitrary $\tau \in I$. This means that

The curve $\alpha(\tau)$ and the vector field $A(\tau)$ unambiguously define the Jacobi variation φ.

The curve $\alpha(\tau)$ and the field $A(\tau)$ are connected with the geodesic $\gamma(t)$ by the relations

$$\alpha(0) = \gamma(t_0), \quad A(0) = \dot{\gamma}(t_0). \tag{1}$$

Obviously, these conditions are not only necessary but also sufficient for the curve $\alpha(\tau)$ and the field $A(\tau)$ to define some Jacobi variation φ of the geodesic γ; that is,

For an arbitrary smooth curve $\alpha(\tau)$ in the space M and an arbitrary vector field $A(\tau)$ on the curve $\alpha(\tau)$ that are connected with the geodesic $\gamma(t)$ by relations (1), there exists a Jacobi variation φ of the geodesic γ such that

$$\alpha(\tau) = \varphi(t_0, \tau), \quad A(\tau) = \frac{\partial \varphi}{\partial t}(t_0, \tau).$$

This variation is defined by the formula

$$\varphi(t, \tau) = \gamma_{\alpha(\tau), A(\tau)}(t),$$

where $\gamma_{\alpha(\tau), A(\tau)}(t)$ is the maximal geodesic passing through the point $\alpha(\tau)$ [for $t = t_0$] and having tangent vector $A(\tau)$ at that point.

We shall call a vector field $X(t)$ on a geodesic $\gamma(t)$ a *Jacobi field* if there exists a Jacobi variation φ of the geodesic γ such that

$$X(t) = \frac{\partial \varphi}{\partial \tau}(t, 0) \quad \text{for every} \quad t \in R,$$

that is, if the field $X(t)$ is associated (in the sense of section 1) with this variation φ. In accordance with what was said above,

An arbitrary Jacobi field $X(t)$ on a geodesic $\gamma(t)$ is unambiguously defined by giving the curve $\alpha(\tau)$ and the vector field $A(\tau)$, both satisfying conditions (1).

Consequently,

For any two vectors A, $B \in M_{\gamma(t_0)}$ there exists on the geodesic $\gamma(t)$, a Jacobi field $X(t)$ such that

$$X(t_0) = A, \quad \frac{\nabla X}{dt}(t_0) = B. \tag{2}$$

To construct this field, it is sufficient to construct a curve $\alpha(\tau)$ such that

$$\alpha(0) = \gamma(t_0), \quad \dot{\alpha}(0) = A,$$

and to construct on this curve a vector field $A(\tau)$ such that

$$A(0) = \dot{\gamma}(t_0), \quad \frac{\nabla A}{dt}(0) = B.$$

Let us now show that

An arbitrary Jacobi field X(t) on a geodesic $\gamma(t)$ satisfies the equation

$$\frac{\nabla}{dt}\frac{\nabla}{dt}X(t) + R_{\gamma(t)}(X(t), \dot{\gamma}(t))\dot{\gamma}(t) = 0. \tag{3}$$

Proof: Let $\varphi(t, \tau)$ denote the Jacobi variation with which the field $X(t)$ is associated. If we apply formula (3) of section 3 of Chapter 2 to the vector field $\frac{\partial\varphi}{\partial t}(t, \tau)$ on the surface φ and keep in mind the fact that, in accordance with formula (2) of section 3 of Chapter 2

$$\frac{\nabla}{\partial\tau}\frac{\partial\varphi}{\partial t}(t, \tau) = \frac{\nabla}{\partial t}\frac{\partial\varphi}{\partial\tau}(t, \tau),$$

we obtain the relation

$$\frac{\nabla}{\partial\tau}\frac{\nabla}{\partial t}\frac{\partial\varphi}{\partial t}(t, \tau) - \frac{\nabla}{\partial t}\frac{\nabla}{\partial t}\frac{\partial\varphi}{\partial\tau}(t, \tau) = \\ = R_{\varphi(t,\tau)}\left(\frac{\partial\varphi}{\partial\tau}(t, \tau), \frac{\partial\varphi}{\partial t}(t, \tau)\right)\frac{\partial\varphi}{\partial t}(t, \tau).$$

If we now set $\tau = 0$ and consider the equation

$$\frac{\partial\varphi}{\partial t}(t, 0) = \dot{\gamma}(t),$$

we immediately obtain equation (3) [since the curve $\gamma(t)$ is a geodesic and hence $\frac{\nabla\dot{\gamma}}{\partial t}(t) = 0$].

Let us suppose now that a ***field of frames***, that is, a system of m vector fields $A_1(t), \ldots, A_m(t)$ such that, for arbitrary $t \in R$, the vectors

$$A_1(t), \ldots, A_m(t) \in M_{\gamma(t)}$$

constitute a basis for the space $M_{\gamma(t)}$, is defined on the geodesic $\gamma(t)$. (Such a system can be constructed, for example, by taking an arbitrary basis for the space $M_{\gamma(t_0)}$ and performing a parallel translation of it into all points of the curve $\gamma(t)$.) Let us consider the components $X^i(t)$ of the vector $X(t)$ relative to the basis $A_1(t), \ldots, A_m(t)$. We can then write equation (3) as a system of m ordinary second-order differential equations for the unknown functions $X^1(t), \ldots, X^m(t)$. Therefore, on the basis of the theorem on the uniqueness of the solution of a system of differential equations, every solution

$X(t)$ of equation (3) [and, in particular, every Jacobi field] is uniquely determined by the values at $t=t_0$ of its components $X^i(t)$ and their first derivatives $\frac{dX^i}{dt}(t)$. But, obviously, these values are uniquely determined by conditions (2). Therefore, in view of the arbitrariness of the construction of the curve $\alpha(\tau)$ and the field $A(\tau)$, we have

The Jacobi field X(t) constructed above from the vectors A and B is unambiguously determined by these vectors.

Furthermore,

Every solution of equation (3) is a Jacobi field.

This is true because each such solution $X(t)$ is uniquely determined by the vectors $A=X(t_0)$ and $B=\frac{\nabla X}{dt}(t_0)$ and hence coincides with the Jacobi field constructed for these vectors.

Since equation (3) is linear, it follows that

The set J_γ of all Jacobi fields on the geodesic γ is a 2m-dimensional linear space.

If we take the scalar product of both sides of equation (3) and an arbitrary Jacobi field $Y(t)$, we obtain

$$\left(\frac{\nabla}{dt}\frac{\nabla}{dt}X(t),\ Y(t)\right)+R_{\gamma(t)}(X(t),\ \dot\gamma(t),\ \dot\gamma(t),\ Y(t))=0.$$

Analogously,

$$\left(\frac{\nabla}{dt}\frac{\nabla}{dt}Y(t),\ X(t)\right)+R_{\gamma(t)}(Y(t),\ \dot\gamma(t),\ \dot\gamma(t),\ X(t))=0.$$

But, by virtue of formulas (1)-(3) of section 2 of Chapter 3,

$$R_{\gamma(t)}(Y(t),\ \dot\gamma(t),\ \dot\gamma(t),\ X(t))=R_{\gamma(t)}(X(t),\ \dot\gamma(t),\ \dot\gamma(t),\ Y(t)).$$

Consequently,

For any two Jacobi fields X(t) and Y(t),

$$\left(\frac{\nabla}{dt}\frac{\nabla}{dt}X(t),\ Y(t)\right)=\left(X(t),\ \frac{\nabla}{dt}\frac{\nabla}{dt}Y(t)\right). \tag{4}$$

Of especial interest to us are the Jacobi fields $X(t)$ such that

$$X(t_0)=0.$$

Obviously, the set $\Lambda_\gamma(t_0)$ of all such fields is a linear subspace of the space J_γ. Also,

The mapping

$$X(t)\to\frac{\nabla X}{dt}(t_0)$$

is an isomorphism of the space $\Lambda_\gamma(t_0)$ onto the space $M_\gamma(t_0)$.

In particular,

The dimension of the space $\Lambda_\gamma(t_0)$ is m.

It follows easily from formula (4) and from formula (8) of section 1 of Chapter 3 that, for any two Jacobi fields $X(t)$ and $Y(t)$,

$$\frac{d}{dt}\left[\left(\frac{\nabla}{dt}X(t),\ Y(t)\right)-\left(X(t),\ \frac{\nabla}{dt}Y(t)\right)\right]=0,$$

that is,

$$\left(\frac{\nabla}{dt}X(t),\ Y(t)\right)-\left(X(t),\ \frac{\nabla}{dt}Y(t)\right)=\text{const.}$$

In particular, if $X(t),\ Y(t)\in\Lambda_\gamma(t_0)$, this expression vanishes for $t=t_0$ and hence vanishes for all t. Thus, we have proven that

For arbitrary fields X(t), Y(t) $\in\Lambda_\gamma(t_0)$,

$$\left(\frac{\nabla}{dt}X(t),\ Y(t)\right)=\left(X(t),\ \frac{\nabla}{dt}Y(t)\right). \tag{5}$$

If we now take the scalar product of each side of equation (3) with the field $\dot{\gamma}(t)$ and remember the symmetry properties of the Riemannian tensor (cf. formula (2), section 2, Chapter 3), we obtain the relation

$$\left(\frac{\nabla}{dt}\frac{\nabla}{dt}X(t),\ \dot{\gamma}(t)\right)=0,$$

that is,

$$\frac{d^2}{dt^2}(X(t),\ \dot{\gamma}(t))=0.$$

Consequently,

$$(X(t),\ \dot{\gamma}(t))=at+b,$$

where a and b are constants.

Now, if $X(t)\in\Lambda_\gamma(t_0)\cap\Lambda_\gamma(t_1)$, where $t_0\neq t_1$, we have

$$at_0+b=at_1+b=0,$$

which is possible only if $a=b=0$. Thus,

For an arbitrary field $X(t)\in\Lambda_\gamma(t_0)\cap\Lambda_\gamma(t_1)$,

$$(X(t),\ \dot{\gamma}(t))=0, \tag{6}$$

that is, any such field is "orthogonal" to the geodesic $\gamma(t)$.

4. Conjugate points

Two points $\gamma(t_0)$ and $\gamma(t_1)$ of a geodesic γ [or the corresponding t_0 and t_1 on the real axis] are said to be *conjugate* if there exists on the geodesic $\gamma(t)$ a nonzero Jacobi field $X(t)$ that vanishes at these two points:

$$X(t_0) = X(t_1) = 0,$$

that is, if the subspace $\Lambda_\gamma(t_0) \cap \Lambda_\gamma(t_1)$ of the space J_γ is nonzero. We shall call the dimension

$$\lambda_{t_1}(t_0) = \dim(\Lambda_\gamma(t_0) \cap \Lambda_\gamma(t_1))$$

of this space the ***index*** of the point $\gamma(t_0)$ [or the point t_0] with respect to the point $\gamma(t_1)$ [or the point t_1]. Thus,

A point $\gamma(t_0)$ ***is conjugate to the point*** $\gamma(t_1)$ ***if and only if its index*** $\lambda_{t_1}(t_0)$ ***is nonzero.***

The value of the concept of conjugate points is conditioned in turn by the following assertion:

If two points t_0 ***and*** t_1 ***are not conjugate, then, for arbitrary vectors*** $A \in M_{\gamma(t_0)}$ ***and*** $B \in M_{\gamma(t_1)}$, ***there exists one and only one field*** $X(t) \in J_\gamma$ ***such that***

$$X(t_0) = A, \quad X(t_1) = B.$$

To prove this assertion, let us consider the linear mapping $\Lambda_\gamma(t_1) \to M_{\gamma(t_0)}$ that assigns to each field $X(t) \in \Lambda_\gamma(t_1)$ the vector $X(t_0) \in M_{\gamma(t_0)}$. The fact that the points t_0 and t_1 are not conjugate means that this mapping is monomorphic and hence isomorphic (since the dimensions of the spaces $\Lambda_\gamma(t_1)$ and $M_{\gamma(t_0)}$ are the same). Consequently, for an arbitrary vector $C \in M_{\gamma(t_0)}$, there exists one and only one field $Y(t) \in \Lambda_\gamma(t_1)$ for which $Y(t_0) = C$.

Now, let $X_0(t)$ denote an arbitrary field in J_γ for which $X_0(t_1) = B$. Then, any other field $X(t) \in J_\gamma$ with this property is defined by the formula

$$X(t) = X_0(t) + Y(t),$$

where $Y(t)$ is some (uniquely defined) field in $\Lambda_\gamma(t_1)$. If we now choose the field $Y(t)$ in such a way that

$$Y(t_0) = A - X_0(t_0),$$

we shall obviously have the desired field $X(t)$.

To study the properties of conjugate points, we note that, when we are constructing fields in $\Lambda_\gamma(t_0)$ with the aid of Jacobi variations, we can, by using the arbitrariness in the choice of the curve $\alpha(\tau)$, assume that the curve $\alpha(\tau)$ is degenerate. Thus,

Every field $X(t) \in \Lambda_\gamma(t_0)$ ***is associated with a Jacobi variation*** $\varphi(t, \tau)$ ***such that***

$$\varphi(t_0, \tau) = \gamma(t_0) \qquad \textit{for all } \tau.$$

In normal coordinates at the point $p_0 = \gamma(t_0)$, this variation $\varphi(t, \tau)$ is given (see end of section 7, Chapter 2) by the functions

$$x^i(t, \tau) = a^i(\tau)(t - t_0), \qquad i = 1, \ldots, m$$

which are linear in t; hence the components $X^i(t)$ of the field $X(t)$ relative to the basis

$$\left(\frac{\partial}{\partial x^1}\right)_{\gamma(t)}, \ldots, \left(\frac{\partial}{\partial x^m}\right)_{\gamma(t)}$$

are also linear with respect to t:

$$X^i(t) = \dot{a}^i(0)(t - t_0), \qquad i = 1, \ldots, m,$$

Therefore,

If the field $X(t) \in \Lambda_\gamma(t_0)$ is not identically zero, it is not equal to zero for any t for which the point $\gamma(t)$ [together with the entire segment $\gamma|_{[t_0, t]}$ of the geodesic γ] belongs to a normal neighborhood of the point $\gamma(t_0)$.

Since an arbitrary point in the space M has a neighborhood that is a normal neighborhood of each of its points, it follows that

Every point on the real axis has a neighborhood any two points in which are nonconjugate.

Therefore,

For an arbitrary finite interval [a, b] of the real axis, there exists a positive number d such that the inequality $|t_1 - t_0| < d$ implies that the points t_0 and t_1 in the interval [a, b] are nonconjugate.

Since the geodesic $\gamma(t)$ between any two of its points that are sufficiently close to each other is a minimizing curve and since the parameter t on it is proportional to the arc length, it follows that

For any two points p_0 and q_0 of a geodesic γ, there exists a positive number d_1 such that the inequality $\rho(p, q) < d_1$ implies that any two points p and q of the geodesic γ, both lying between the points p_0 and q_0, are nonconjugate.

In the case in which the space M is compact and hence there exists a positive number d such that any two points $p, q \in M$ for which $\rho(p, q) < d$ belong to a single normal convex neighborhood V, it follows immediately from what has been said that

Points $p, q \in M$, where $\rho(p, q) < d$, are nonconjugate on the minimizing curve $\gamma_{p,q}$ connecting them.

Let us now find all points t_0 that are conjugate [with respect to a given geodesic $\gamma(t)$] to some fixed points $t_1 \in R$.

Let $X_1(t), \ldots, X_m(t)$ denote an arbitrary basis in the space $\Lambda_\gamma(t_1)$. If we choose on the geodesic $\gamma(t)$ the field of frames $A_1(t), \ldots, A_m(t)$ [see above, section 3], we obtain the decompositions

$$X_i(t) = X_i^j(t) A_j(t), \qquad i, j = 1, \ldots, m,$$

where the $X_i^j(t)$ are smooth functions.

By definitions, the point t_0 is conjugate to the point t_1 if and only if there exists a nontrivial linear combination of fields $X_1(t), \ldots, X_m(t)$ with constant coefficients that vanishes at the point t_0. Therefore,

A point t_0 *is conjugate to a point* t_1 *if and only if the determinant*

$$D(t) = \det \left| X_i^j(t) \right|$$

vanishes at $t = t_0$.

It is obvious that (up to a nonzero constant factor) the determinant $D(t)$ is independent of the choice of basis $X_1(t), \ldots, X_m(t)$. Using this fact (and assuming that the points t_0 and t_1 are conjugate), we choose this basis in such a way that the first $\lambda = \lambda_{t_1}(t_0)$ of its fields will constitute a basis for the space $\Lambda_\gamma(t_0) \cap \Lambda_\gamma(t_1)$. Then, $X_i^j(t_0) = 0$ for $1 \leqslant i \leqslant \lambda$ and $1 \leqslant j \leqslant m$. Therefore, close to the point t_0 (that is, on some interval with center at that point), we have

$$X_i^j(t) = (t - t_0)\,\varphi_i^j(t), \quad 1 \leqslant i \leqslant \lambda, \quad 1 \leqslant j \leqslant m,$$

where the $\varphi_i^j(t)$ are continuous functions. It follows from this that

Close to the point t_0,

$$D(t) = (t - t_0)^\lambda\, \Phi(t),$$

where $\Phi(t)$ *is a continuous function.*

Here, it is easy to see that

$$\Phi(t_0) \neq 0. \tag{1}$$

This is true because

$$\varphi_i^j(t_0) = \frac{dX_i^j}{dt}(t_0) = \frac{\nabla X_i^j}{dt}(t_0),$$

where $\dfrac{\nabla X_i^j}{dt}(t_0)$ is the jth component of the vector $\dfrac{\nabla X_i}{dt}(t_0)$ and hence

$$\Phi(t_0) = \begin{vmatrix} \dfrac{\nabla X_1^1}{dt}(t_0), & \ldots, & \dfrac{\nabla X_1^m}{dt}(t_0) \\ \cdot & \cdots & \cdot \\ \dfrac{\nabla X_\lambda^1}{dt}(t_0), & \ldots, & \dfrac{\nabla X_\lambda^m}{dt}(t_0) \\ X_{\lambda+1}^1(t_0), & \ldots, & X_{\lambda+1}^m(t_0) \\ \cdot & \cdots & \cdot \\ X_m^1(t_0), & \ldots, & X_m^m(t_0) \end{vmatrix}.$$

Therefore, if $\Phi(t_0) = 0$, there exist numbers $c^1, \ldots, c^m$, not all zero, such that, for the fields

$$U(t) = c^1 X_1(t) + \ldots + c^\lambda X_\lambda(t),$$
$$V(t) = c^{\lambda+1} X_{\lambda+1}(t) + \ldots + c^m X_m(t)$$

(if $\lambda = m$, then $V(t) = 0$), we have

$$\frac{\nabla U}{dt}(t_0) = V(t_0).$$

Since $U(t), V(t) \in \Lambda_\gamma(t_1)$, we have, in accordance with formula (5) of section 3,

$$\left(\frac{\nabla U}{dt}(t), V(t)\right) = \left(U(t), \frac{\nabla V}{dt}(t)\right),$$

and hence

$$(V(t_0), V(t_0)) = \left(\frac{\nabla U}{dt}(t_0), V(t_0)\right) = \left(U(t_0), \frac{\nabla V}{dt}(t_0)\right) = 0,$$

because, by construction, $U(t) \in \Lambda_\gamma(t_0)$. Consequently, $V(t_0) = 0$; that is, $V(t) \in \Lambda_\gamma(t_0)$, which is possible only if the field $V(t)$ is identically equal to zero. Thus,

$$c^{\lambda+1} = \ldots = c^m = 0.$$

Furthermore,

$$\frac{\nabla U}{dt}(t_0) = V(t_0) = 0,$$

since $U(t) \in \Lambda_\gamma(t_0)$, it follows from this last equation that $U(t) = 0$, that is, that

$$c^1 = \ldots = c^\lambda = 0.$$

Thus, in contradiction with the assumption, all the numbers $c^1, \ldots, c^m$ are equal to zero. This contradiction proves inequality (1).

Because of the continuity of the function $\Phi(t)$, it follows immediately from inequality (1) that

Close to the point t_0, the determinant $D(t)$ vanishes only at the point t_0; that is, close to the point t_0, there is no point $t \neq t_0$ conjugate to the point t_1.

Since, as we have shown, this assertion is also valid for $t_0 = t_1$, it now follows that

Points conjugate to the point t_1 are discrete; that is, the set of such points has no cluster points in R.

In particular,

On every finite (open or closed) interval I of the real axis, there exist only finitely many points conjugate to the point t_1.

We shall call the sum of the indices of these points the index of the interval I and shall denote it by $\lambda_{t_1}(I)$.

Remark: The fact that there are no points conjugate to the point t_1 close to the point t_1 can also be proved without using the Jacobian variations. One can easily see that, close to the point t_1,

$$D(t) = (t - t_1)^m \Psi(t),$$

where

$$\Psi(t_1) = \det\left|\frac{\nabla X_i^j}{dt}(t_1)\right| \neq 0.$$

5. Piecewise-smooth and discontinuous vector fields

For a more profound study of Jacobi fields and conjugate points, we shall find it convenient to generalize somewhat the concept of a vector field on a curve.

Let $\gamma(t)$, where $a \leqslant t \leqslant b$, denote an arbitrary regular curve (more precisely, a segment of such a curve; in the applications of the general theory that we are about to develop, the curve γ will be a geodesic). We shall say that a function $X(t)$ defined on the interval $[a, b]$ into the space $M_{\gamma(t)}$ is a piecewise-smooth vector field on the curve $\gamma(t)$ if there exists a decomposition

$$a = a_0 < a_1 < \ldots < a_n < a_{n+1} = b$$

of the interval [a, b] such that on each subinterval $[a_i, a_{i+1}]$, the function $X(t)$ is a (smooth) vector field. Let us denote the field $X(t)$ on the interval $[a_i, a_{i+1}]$ by $X_i(t)$. We shall call the points $a_1, \ldots, a_n$ break points of the field $X(t)$. We emphasize that these points depend in general on the field $X(t)$. (In particular, their number n may vary from field to field.)

Obviously, we can introduce the structure of a linear space in a natural manner into the set of all piecewise-smooth fields on the curve $\gamma(t)$, for $a \leqslant t \leqslant b$. We shall denote this space by the symbol $\mathfrak{M}_\gamma$ or simply $\mathfrak{M}$.

Together with the space $\mathfrak{M}$, we shall also need the space $\mathfrak{M}'$ consisting of discontinuous fields $X(t)$, that is, of functions $X(t) \in M_{\gamma(t)}$ that are defined on the interval $[a, b]$ except at a finite number of points of discontinuity $a_1 < \ldots < a_n$ and that satisfy the smoothness condition away from these points. Each such field $X(t)$ can be regarded as a set of smooth fields $X_i(t)$ defined respectively on the intervals $[a_i, a_{i+1}]$. We set

$$X(a_i - 0) = X_{i-1}(a_i), \quad X(a_i + 0) = X_i(a_i).$$

If $X(a_i - 0) = X(a_i + 0)$, we define the field $X(t)$ at the point a_i by

$$X(a_i) = X(a_i - 0) = X(a_i + 0).$$

Under this convention, the space $\mathfrak{M}$ will obviously be a subspace of the space $\mathfrak{M}'$.

For every field $X(t) \in \mathfrak{M}$ (and even for every field $X(t) \in \mathfrak{M}'$), the vectors $\frac{\nabla X}{dt}(t)$, for $t \neq a_1, \ldots, a_n$, constitute a vector field $\frac{\nabla X}{dt}(t) \in \mathfrak{M}'$. If this field belongs to the space $\mathfrak{M}$, that is, if, for arbitrary $i = 1, \ldots, n$,

$$\frac{\nabla X}{dt}(a_i - 0) = \frac{\nabla X}{dt}(a_i + 0),$$

we shall say that the field $X(t) \in \mathfrak{M}$ is differentiable. Analogously, we shall say that a differentiable field $X(t)$ for which the field $\frac{\nabla}{dt}\frac{\nabla}{dt}X(t)$ belongs to the space $\mathfrak{M}$ is a twice differentiable field, etc.

Twice differentiable fields are not in general smooth fields.

If a twice differentiable field $X(t)$ satisfies equation (3) of section 3, it is smooth; that is, it is a Jacobi field.

Proof (Cf. analogous reasoning for geodesics in section 6 of Chapter 2): If we differentiate equation (3) of section 3 covariantly, we can express all the higher covariant derivatives of the field $X(t)$ successively in terms of the fields $X(t)$, $\frac{\nabla X}{dt}(t)$, and $\frac{\nabla}{dt}\frac{\nabla}{dt}X(t)$. Consequently, all the higher covariant derivatives exist for this field, from which it follows immediately that it is a smooth field.

We shall say that a smooth field $Y(t) \in \mathfrak{M}$ is a *covariant integral* of the field $X(t) \in \mathfrak{M}'$ if

$$\frac{\nabla Y}{dt}(t) = X(t)$$

for all t for which the field $X(t)$ is defined.

Obviously, if this field $X(t)$ is piecewise-smooth, the field $Y(t)$ is differentiable.

Furthermore, it is easy to see that

For an arbitrary field $X(t) \in \mathfrak{M}'$ and an arbitrary vector $A \in M_{\gamma(a)}$, there exists one and only one field $Y(t) \in \mathfrak{M}$ that is a covariant integral of the field $X(t)$ and that satisfies the condition $Y(a) = A$.

The construction of this field on every interval of the curve $\gamma(t)$ not containing points of discontinuity of the field $X(t)$ reduces in an obvious way to solving a nonhomogeneous system of first-order differential equations. At points of discontinuity, these solutions must be adjusted to fit each other.

One can also prove the existence of a covariant integral $Y(t)$ without using differential equations. First of all, it is obviously sufficient to prove the existence of a covariant integral $Y_0(t)$ that vanishes at the point a because, as one can easily see, the covariant integral $Y_0(t)$ such that $Y(a) = A$ is expressed in terms of the integral $Y_0(t)$ in accordance with the formula

$$Y(t) = Y_0(t) + \tau_a^t A,$$

where $\tau_a^t : M_{\gamma(a)} \to M_{\gamma(t)}$ is a parallel translation along the curve γ that we are considering.

Let us now consider an arbitrary partition

$$a = t_0 < t_1 < \ldots < t_n < t_{n+1} = t$$

of the interval $[a, t]$ and let us assign to it the approximating sum

$$\sum_{i=0}^{n} \tau_{t_i}^t X(t_i)(t_{i+1} - t_i),$$

where $\tau_{t_i}^t : M_{\gamma(t_i)} \to M_{\gamma(t)}$ is a parallel translation along the curve γ. The limit (which obviously exists) of approximating sums of this kind as $\max_i |t_{i+1} - t_i| \to 0$ is analogous to the definite integral

$\int_{a}^{t} f(t)\,dt$ with variable upper limit of integration, with which we are familiar from calculus. Repeating step by step the reasoning followed in calculus, we see that this limit is the covariant integral $Y_0(t)$.

Obviously, the properties of the covariant integral $Y_0(t)$ are analogous to the properties of an ordinary integral $\int_{a}^{t} f(t)\,dt$. We shall not enumerate all these properties (since we shall not need all) but shall merely note the following property (which we shall have occasion to use):

For every $t \in [a, b]$, *every field* $X(t) \in \mathfrak{M}$, *and every vector* $B_t \in M_{\gamma(t)}$,

$$(B_t, \, Y_0(t)) = \int_{0}^{t} (B_t, \, \tau_\theta^t X(\theta))\,d\theta, \tag{1}$$

where $Y_0(t)$ *is the covariant integral of the field* $X(t)$ *that vanishes at the point* a *and* $\tau_\theta^t : M_{\gamma(\theta)} \to M_{\gamma(t)}$ *is a parallel translation along the curve* γ.

To prove this formula, it suffices to note that, by definition, its left- and right-hand members are limits of the same sum

$$\left(B_t, \, \sum_{i=0}^{n} \tau_{t_i}^t X(t_i)(t_{i+1} - t_i) \right) = \sum_{i=0}^{n} (B_t, \, \tau_{t_i}' X(t_i))(t_{i+1} - t_i).$$

In conclusion, let us look at an arbitrary family $X_\theta(t)$ of fields in $\mathfrak{M}'$ that depend on a parameter $\theta \in (-\theta_0, \theta_0)$, where θ_0 is some positive number.

We shall say that fields in this family depend smoothly on the parameter θ if, for arbitrary $t \in [a, b]$ for which the vector $X_\theta(t)$ is defined, its components $X_\theta^i(t)$ in some system of local coordinates $x^1, \ldots, x^m$ at the point $\gamma(t)$ are smooth functions of the parameter θ. Obviously, this condition does not depend on the choice of local coordinates.

Let $\partial X_\theta^i(t)/\partial\theta$ denote the partial derivatives of the components $X_\theta^i(t)$ with respect to the parameter θ. It is easy to see that the derivative

$$\frac{\partial X_\theta(t)}{\partial\theta} = \frac{\partial X_\theta^i}{\partial\theta} \left(\frac{\partial}{\partial x^i} \right)_{\gamma(t)}$$

is independent of the choice of coordinates $x^1, \ldots, x^m$ and is uniquely determined by the family $X_\theta(t)$. The field $\partial X_\theta(t)/\partial\theta$ obtained also belongs to the space $\mathfrak{M}'$ and depends smoothly on the parameter θ. If the fields $X_\theta(t)$ are piecewise smooth (that is, belong to the space $\mathfrak{M}$), then the field $\partial X_0(t)/\partial\theta$ also belongs to the space $\mathfrak{M}$.

In what follows, we shall use the following simple proposition, the proof of which we leave to the reader:

For arbitrary fields X(t), $A(t) \in \mathfrak{M}$, *there exists a family* $X_\theta(t) \in \mathfrak{M}$ *that depends smoothly on the parameter* θ *and that satisfies the conditions*

$$X_0(t) = X(t), \quad \left.\frac{\partial X_\theta(t)}{\partial \theta}\right|_{\theta=0} = A(t).$$

If the field A(t) vanishes at the points A and B, then the family $X_\theta(t)$ *can be constructed so that* $X_\theta(a) = X(a)$ *and* $X_\theta(b) = X(b)$ *for arbitrary* θ.

6. Minimal vector fields

Consider now the quadratic functional

$$I_a^b(X(t)) = \int_a^b \left[\left(\frac{\nabla X}{dt}(t), \frac{\nabla X}{dt}(t)\right) - R_{\gamma(t)}(X(t), \dot{\gamma}(t), \dot{\gamma}(t), X(t))\right] dt,$$

defined on the space $\mathfrak{M} = \mathfrak{M}_\gamma$. (The fact that the integrand is not defined at individual points in the interval $[a, b]$ does not keep the integral from existing.) We shall assume that the curve $\gamma(t)$, on which the fields $X(t) \in \mathfrak{M}$ are being considered, is a geodesic.

We shall say that field $X(t) \in \mathfrak{M}$ is *minimal* if

$$I_a^b(X(t)) \leqslant I_a^b(Y(t))$$

for an arbitrary field $Y(t) \in \mathfrak{M}$ that coincides at the end-points of the interval $[a, b]$ with the field $X(t)$, that is, for a field $Y(t)$ such that

$$Y(a) = X(a), \quad Y(b) = X(b).$$

Assuming that minimal fields exist, let us consider an arbitrary such field $X(t)$. As we know (cf. end of section 5), for an arbitrary piecewise-smooth field $A(t)$ on a geodesic $\gamma(t)$ that vanishes at the points a and b, there exists a family $X_\theta(t)$ of piecewise-smooth fields depending smoothly on the parameter $\theta \in (-\theta_0, \theta_0)$ and possessing the property that

$$X_0(t) = X(t), \quad X_\theta(a) = X(a), \quad X_\theta(b) = X(b),$$
$$\left.\frac{\partial X_\theta(t)}{\partial \theta}\right|_{\theta=0} = A(t).$$

Let us consider the function

$$I(\theta) = I_a^b(X_\theta(t)).$$

If we differentiate this function with respect to θ, we obtain (using, in particular, the symmetry properties of the Riemannian tensor)

$$I'(\theta)=2\int_a^b\Big[\Big(\frac{\nabla X_\theta}{dt}(t),\ \frac{\partial}{\partial\theta}\frac{\nabla X_\theta}{dt}(t)\Big)-$$
$$-R_{\gamma(t)}\Big(X_\theta(t),\ \dot\gamma(t),\ \dot\gamma(t),\ \frac{\partial X_\theta(t)}{\partial t}\Big)\Big]\,dt$$

and, consequently,

$$I'(0)=2\int_a^b\Big[\Big(\frac{\nabla X}{dt}(t),\ \frac{\nabla A}{dt}(t)\Big)-$$
$$-R_{\gamma(t)}(X(t),\ \dot\gamma(t),\ \dot\gamma(t),\ A(t))\Big]\,dt.$$

Let $Y_0(t)$ denote the covariant integral of the field

$$-R_{\gamma(t)}(X(t),\ \dot\gamma(t))\,\dot\gamma(t),$$

that vanishes at the point a. In accordance with formula (8) of section 1 of Chapter 3 (which is obviously valid also for piecewise-smooth fields away from their break points), we have

$$-R_{\gamma(t)}(X(t),\ \dot\gamma(t),\dot\gamma(t),\ A(t))=$$
$$=(-R_{\gamma(t)}(X(t),\ \dot\gamma(t))\,\dot\gamma(t),\ A(t))=\Big(\frac{\nabla Y_0}{dt}(t),\ A(t)\Big)=$$
$$=\frac{d}{dt}(Y_0(t),\ A(t))-\Big(Y_0(t),\ \frac{\nabla A}{dt}(t)\Big).$$

Since, by hypothesis, $A(a)=0$ and $A(b)=0$, it follows that

$$I'(0)=2\int_a^b\Big(\frac{\nabla X}{dt}(t)-Y_0(t),\ \frac{\nabla A}{dt}(t)\Big)\,dt. \qquad (1)$$

But, because the field $X(t)$ is a minimal field, the function $I(\theta)$ has, for an arbitrary family $X_\theta(t)$, a minimum at the point $\theta = 0$, and hence $I'(0) = 0$. This proves that

If a field X(t) is minimal, then for an arbitrary field $A(t)\in\mathfrak{M}$ *that vanishes at the points a and b,*

$$\int_a^b\Big(\frac{\nabla X}{dt}(t)-Y_0(t),\ \frac{\nabla A}{dt}(t)\Big)\,dt=0,$$

where $Y_0(t)$ *is the covariant integral of the field*

$$-R_{\gamma(t)}(X(t),\ \dot\gamma(t))\,\dot\gamma(t),$$

that vanishes at the point a.

Now, let $Z_0(t)$ denote the covariant integral of the field $\frac{\nabla X}{dt}(t) - Y_0(t)$ that vanishes at the point a and let $B(t)$ denote the vector field on the geodesic $\gamma(t)$ obtained by parallel translation of the vector $Z_0(b)/b-a$ along the curve $\gamma(t)$. Consider the field

$$A(t) = Z_0(t) - (t-a)B(t).$$

Obviously, $A(a)=0$ and $A(b)=0$. Therefore,

$$\int_a^b \left(\frac{\nabla X}{dt}(t) - Y_0(t),\ \frac{\nabla A}{dt}(t)\right) dt = 0,$$

But,

$$\frac{\nabla A}{dt}(t) = \frac{\nabla Z_0}{dt}(t) - B(t) - (t-a)\frac{\nabla B}{dt}(t) = $$
$$= \frac{\nabla X}{dt}(t) - Y_0(t) - B(t),$$

since $\frac{\nabla B}{dt}(t)=0$. Consequently,

$$\int_a^b \left(\frac{\nabla X}{dt}(t) - Y_0(t),\ \frac{\nabla X}{dt}(t) - Y_0(t) - B(t)\right) dt = 0,$$

and, therefore,

$$\int_a^b \left(\frac{\nabla X}{dt}(t) - Y_0(t) - B(t),\ \frac{\nabla X}{dt}(t) - Y_0(t) - B(t)\right) dt =$$
$$= -\int_a^b \left(B(t),\ \frac{\nabla X}{dt}(t) - Y_0(t) - B(t)\right) dt =$$
$$= -\int_a^b \left(B(t),\ \frac{\nabla Z_0}{dt}(t) - B(t)\right) dt =$$
$$= -\int_a^b \frac{d}{dt}(B(t),\ Z_0(t))\, dt + \int_a^b (B(t),\ B(t))\, dt =$$
$$= -(B(t),\ Z_0(t))\Big|_a^b + (B(t),\ B(t))(b-a) =$$
$$= -(B(b),\ Z_0(b)) + (B(b),\ B(b))(b-a) = 0.$$

(Here, we use the fact that the scalar product $(B(t),\ B(t))$ is independent of t and, in particular, is equal to $(B(b),\ B(b))$). This proves that

$$\frac{\nabla X}{dt}(t) - Y_0(t) - B(t) = 0,$$

that is, that

$$\frac{\nabla X}{dt}(t) = Y_0(t) + B(t). \tag{2}$$

Since the fields $Y_0(t)$ and $B(t)$ are differentiable and since

$$\frac{\nabla Y_0}{dt}(t) = -R(X(t), \dot{\gamma}(t))\dot{\gamma}(t) \text{ and } \frac{\nabla B}{dt}(t) = 0,$$

it follows immediately that the field $X(t)$ is twice differentiable and satisfies equation (3) of section 3. Consequently, this field is smooth and hence is a Jacobi field. In other words, we have shown that

Every minimal field is a Jacobi field.

7. The existence of minimal fields

The results proven in the preceding section are of a conditional nature inasmuch as the question of the existence of minimal fields still remains open. In the present section, we shall prove that

If the interval $[a, b]$ contains no points conjugate with the point b, then, for an arbitrary field $Y(t) \in \mathfrak{M}$,

$$I_a^b(X(t)) \leqslant I_a^b(Y(t)), \tag{1}$$

where $X(t)$ is a Jacobi field that assumes the same values at the end-points of the interval $[a, b]$ as does the field $Y(t)$:

$$X(a) = Y(a), \quad X(b) = Y(b);$$

here, equality holds in formula (1) only when $Y(t) = X(t)$.

In other words,

With the restriction indicated on the interval $[a, b]$, an arbitrary Jacobi field $X(t)$ is minimal.

This theorem completely answers the question of the existence (and uniqueness) of minimal fields inasmuch as, for a not conjugate to b, there exists, as we know, for arbitrary vectors $A \in M_{\gamma(a)}$ and $B \in M_{\gamma(b)}$ a unique Jacobi field $X(t)$ such that

$$X(a) = A, \quad X(b) = B.$$

Let $X_1(t), \ldots, X_m(t)$ denote an arbitrary basis for the space $\Lambda_\gamma(b)$. Let us assign to an arbitrary point $z = (z^i)$ of the space R^m the field $X_z(t) \in J_\gamma$ defined by

$$X_z(t) = X(t) + z^i X_i(t),$$

and let us consider in the strip $[a, b] \times R^m$ of the space R^{m+1} the line integral (the so-called "Hilbert integral")

$$G = \int_C \Big\{ \Big[\Big(\frac{\nabla X_z}{dt}(t), \frac{\nabla X_z}{dt}(t) \Big) - \\ - R_{\gamma(t)}(X_z(t), \dot{\gamma}(t), \dot{\gamma}(t), X_z(t)) \Big] dt + \\ + 2 \Big(\frac{\nabla X_z}{dt}(t), X_i(t) \Big) dz^i \Big\},$$

where C is an arbitrary curve lying in that strip.

It turns out that

The integral G between any two points in the strip $[a, b] \times R^m$ is independent of the path of integration; that is, its integrand is an exact differential.

Proof: For arbitrary $i = 1, \ldots, m$, we have

$$\frac{\partial}{\partial z^i} \Big[\Big(\frac{\nabla X_z}{dt}(t), \frac{\nabla X_z}{dt}(t) \Big) - R_{\gamma(t)}(X_z(t), \dot{\gamma}(t), \dot{\gamma}(t), X_z(t)) \Big] = \\ = 2 \Big[\Big(\frac{\nabla X_z}{dt}(t), \frac{\partial}{\partial z^i} \frac{\nabla X_z}{dt}(t) \Big) - R_{\gamma(t)}\Big(X_z(t), \dot{\gamma}(t), \dot{\gamma}(t), \frac{\partial X_z}{\partial z^i}(t)\Big) \Big] = \\ = 2 \Big[\Big(\frac{\nabla X_z}{dt}(t), \frac{\nabla X_i}{dt}(t) \Big) - R_{\gamma(t)}(X_z(t), \dot{\gamma}(t), \dot{\gamma}(t), X_i(t)) \Big] = \\ = 2 \Big[\Big(\frac{\nabla X_z}{dt}(t), \frac{\nabla X_i}{dt}(t) \Big) + \Big(\frac{\nabla}{dt} \frac{\nabla}{dt} X_z(t), X_i(t) \Big) \Big] = \\ = \frac{\partial}{\partial t} \Big[2 \Big(\frac{\nabla X_z}{dt}(t), X_i(t) \Big) \Big],$$

and, for arbitrary $i, j = 1, \ldots, m$, the expression

$$\frac{\partial}{\partial z^j} \Big[2 \Big(\frac{\nabla X_z}{dt}(t), X_i(t) \Big) \Big] = 2 \Big(\frac{\partial}{\partial z^j} \frac{\nabla X_z}{dt}(t), X_i(t) \Big) = \\ = 2 \Big(\frac{\nabla X_j}{dt}(t), X_i(t) \Big)$$

is, by virtue of formula (5) of section 3, symmetric in i and j.

In particular, let us consider in the strip $[a, b] \times R^m$ a curve C with parametric equations

$$t = t, \\ z^i = z^i(t), \quad i = 1, \ldots, m,$$

where the $z^i(t)$ are arbitrary piecewise-smooth functions on $[a, b]$ that vanish at the points a and b. From what was proven above, the integral G assumes the same value G_0 on all such curves.

On the other hand, each curve defines on $\gamma(t)$ a piecewise-smooth field

$$Y(t) = X(t) + z^i(t) X_i(t),$$

for which $Y(a) = A$ and $Y(b) = B$. Here, since

$$\frac{\nabla Y(t)}{dt} = \frac{\nabla X_z(t)}{dt} \Big|_{z^i = z^i(t)} + \dot{z}^i(t) X_i(t),$$

we have

$$G_0 = \int_a^b \Big[\Big(\frac{\nabla X_z(t)}{dt}, \frac{\nabla X_z(t)}{dt} \Big) \Big|_{z^i = z^i(t)} - - R_{\gamma(t)}(Y(t), \dot{\gamma}(t), \dot{\gamma}(t), Y(t)) + + 2 \Big(\frac{\nabla X_z(t)}{dt} \Big|_{z^i = z^i(t)}, \frac{\nabla Y}{dt}(t) - \frac{\nabla X_z(t)}{dt} \Big|_{z^i = z^i(t)} \Big) \Big] dt$$

and, consequently,

$$I_a^b(Y(t)) - G_0 = = \int_a^b \Big[\Big(\frac{\nabla Y}{dt}(t), \frac{\nabla Y}{dt}(t) \Big) + \Big(\frac{\nabla X_z}{dt}(t), \frac{\nabla X_z}{dt}(t) \Big) \Big|_{z^i = z^i(t)} - - 2 \Big(\frac{\nabla X_z(t)}{dt} \Big|_{z^i = z^i(t)}, \frac{\nabla Y}{dt}(t) \Big) \Big] dt = = \int_a^b \Big(\frac{\nabla Y}{dt}(t) - \frac{\nabla X_z(t)}{dt} \Big|_{z^i = z^i(t)}, \frac{\nabla Y}{dt}(t) - \frac{\nabla X_z(t)}{dt} \Big|_{z^i = z^i(t)} \Big) dt \geqslant 0,$$

Equality holds in this relation only when

$$\frac{\nabla Y}{dt}(t) = \frac{\nabla X_z(t)}{dt} \Big|_{z^i = z^i(t)},$$

that is, when the $z^i(t) \equiv 0$ (so that, in particular, $G_0 = I_a^b(X(t))$). This proves that, for an arbitrary field $Y(t)$ of the form $X(t) + z^i(t) X_i(t)$,

$$I_a^b(X(t)) \leqslant I_a^b(Y(t)),$$

with equality holding only when $Y(t) = X(t)$.

To complete the proof, it remains to note that any field $Y(t) \in \mathfrak{M}$ that assumes values A and B at the points a and b, respectively, can be represented (in a natural manner) in the form $X(t) + z^i(t) X_i(t)$ since, for an arbitrary point $t \in (a, b)$, the vectors $X_1(t), \ldots, X_m(t)$ constitute a basis for the space $M_{\gamma(t)}$. (Otherwise, some nontrivial linear combination of them would be equal to 0 and therefore the point t would, in contradiction with the hypothesis, be conjugate to the point b.)

Since the field $X(t) \equiv 0$ is the unique Jacobi field that vanishes at the points a and b when a is not conjugate to b it follows from the assertion just proven that

If the interval [a, b] does not contain points conjugate to the point b, then, for an arbitrary nonzero field $X(t) \in \mathfrak{M}$ that vanishes at the points a and b,

$$I_a^b(X(t)) > 0. \qquad (2)$$

In connection with the results of this and the preceding section, the question naturally arises as to the evaluation of the functional I_a^b on an arbitrary Jacobi field. We shall show that

For an arbitrary Jacobi field X(t),

$$I_a^b(X(t)) = \left(\frac{\nabla X}{dt}(b),\ X(b)\right) - \left(\frac{\nabla X}{dt}(a),\ X(a)\right). \tag{3}$$

Proof: If we multiply equation (3) of section 3 by $X(t)$ and integrate with respect to t from a to b, we obtain

$$\int_a^b \left(\frac{\nabla}{dt}\frac{\nabla}{dt}X(t),\ X(t)\right)dt = -\int_a^b R_{\gamma(t)}(X(t),\ \dot{\gamma}(t),\ \dot{\gamma}(t),\ X(t))\,dt.$$

Consequently,

$$I_a^b(X(t)) = \int_a^b \left[\left(\frac{\nabla X}{dt}(t),\ \frac{\nabla X}{dt}(t)\right) + \left(\frac{\nabla}{dt}\frac{\nabla}{dt}X(t),\ X(t)\right)\right]dt =$$

$$= \int_a^b \frac{d}{dt}\left(\frac{\nabla X}{dt}(t),\ X(t)\right)dt = \left(\frac{\nabla X}{dt}(t),\ X(t)\right)\Big|_a^b.$$

8. Broken Jacobi fields

Let $\bar{p}$ and $\bar{q}$ denote any two points on a geodesic $\gamma(t)$. To make the formulas as simple as possible, we shall assume that the parameter t on the geodesic $\gamma(t)$ is chosen in such a way that

$$\bar{p} = \gamma(0), \quad \bar{q} = \gamma(1).$$

Suppose also that

$$0 \leqslant \theta < 1$$

and that

$$\theta = t_0 < t_1 < \ldots < t_s < t_{s+1} = 1. \tag{1}$$

We shall say that a piecewise-smooth field $X(t)$ defined on a segment $\gamma|_{[\theta,\,1]}$ of the geodesic γ is a ***broken Jacobi field*** (with breaks at the points $t_1, \ldots, t_s$) if every field

$$X_i(t) = X(t)|_{[t_i,\, t_{i+1}]}, \quad i = 0, \ldots, s$$

is a Jacobi field on the segment $\gamma|_{[t_i,\, t_{i+1}]}$.

Let Θ denote the set of all broken Jacobi fields $X(t)$ on the segment $\gamma|_{[\theta, 1]}$ that vanish at the end-points of that segment:

$$X(\theta) = 0, \quad X(1) = 0,$$

that have breaks at the given points $t_1, \ldots, t_s$, and that are orthogonal to the geodesic $\gamma(t)$, that is, that have the property that

$$(X(t), \dot{\gamma}(t)) = 0 \quad \text{for } \theta \leqslant t \leqslant 1.$$

Obviously, the structure of a linear space can be introduced into this set in a natural manner.

In the present section, we shall describe a certain procedure that will enable us to construct all fields in Θ and, in particular, to find the dimension of the space Θ. Here, we shall assume that the numbers (1) are distributed sufficiently "thickly" in the interval $[\theta, 1]$, specifically, in such a way that, for arbitrary $i = 0, \ldots, s$, the points

$$p_i = \gamma(t_i), \quad p_{i+1} = \gamma(t_{i+1})$$

of the geodesic $\gamma(t)$ belong to some normal convex neighborhood $V_i \subset M$. By virtue of this assumption (cf. section 4), we have the following:

For arbitrary $i = 0, \ldots, s$, *the interval* $[t_i, t_{i+1}]$ *has no points conjugate to the point* t_{i+1}; *in particular, the points* t_i *and* t_{i+1} *are not conjugate. Furthermore, there exists a positive number* δ *such that any two points p,* $p, q \in M$ *that satisfy the inequalities*

$$\rho(p_i, p) < \delta, \quad \rho(p_{i+1}, q) < \delta,$$

for some $i = 0, 1, \ldots, s$ *can be connected in M by a unique minimizing curve.*

Proof: Obviously, any positive δ satisfies this condition if, for every $j = 0, 1, \ldots, s+1$, the spherical δ-neighborhood of the point p_j is contained in the intersection $V_{j-1} \cap V_j$ of the neighborhoods V_{j-1} and V_j (for $j = 0$, in the neighborhood V_0 and for $j = s + 1$, in the neighborhood V_s).

In what follows, we shall assume that we have chosen such a positive number δ once and for all so that for no $i = 0, 1, \ldots, s$ do the spherical δ-neighborhoods of the points p_i and p_{i+1} intersect.

Let us suppose that for arbitrary $i = 1, \ldots, s$, we have chosen an $(m - 1)$-dimensional submanifold P_i of the manifold M that is contained in the δ-neighborhood of the point p_i, that passes through the point p_i, and such that the geodesic γ is not tangent to it at the point p_i (that is, its tangent vector $\dot{\gamma}_i = \dot{\gamma}(t_i)$ at that point is not tangent to the submanifold). For example, let f denote an arbitrary smooth function such that $f(p_i) = 0$ and $(df)_{p_i}(\dot{\gamma}_i) \neq 0$ (let us say, of each local coordinate x^j for which $\dot{\gamma}_i x^j \neq 0$). Then that portion of the level surface $[f = 0]$ contained in the δ-neighborhood of the point p_i is such a submanifold. We shall assume that the submanifolds

$$P_1, \ldots, P_s \tag{2}$$

are fixed once and for all.

We shall also assume that the positive number δ is so small that, on every manifold P_i, there exists a system of local coordinates

$$z_i^1, \ldots, z_i^{m-1}, \tag{3}$$

of which the entire manifold P_i is a coordinate neighborhood. We shall assume that the coordinates (3) vanish at the point p_i and that these too are chosen once and for all.

Let U_i denote an open set of the space R^{m-1} and suppose that U_1 is the image of the manifold P_i under the coordinate homeomorphism defined by the coordinates 3. Let

$$U = U_1 \times \ldots \times U_s$$

denote the Cartesian product of the sets $U_1, \ldots, U_s$. The set U is an open submanifold of the $s(m-1)$-dimensional Euclidean space $R^{s(m-1)}$, and each point $z = (z_i^k)$ in it, for $i = 1, \ldots, s$ and $k = 1, \ldots, m-1$ determines the s points $q_i \in P_i$, for $i = 1, \ldots, s$, that are related with the point z by

$$z_i^k(q_i) = z_i^k, \quad k = 1, \ldots, m-1.$$

The points q_i, treated as points in the space M, are such that

$$\rho(p_i, q_i) < \delta, \quad i = 1, \ldots, s.$$

Therefore, for arbitrary $i = 1, \ldots, s-1$, the points q_i and q_{i+1} can be connected in M by a unique minimizing curve. Furthermore, the point $p_0 = \gamma(\theta)$ can be connected by a unique minimizing curve to the point q_1 and the point q_s can be connected to the point $\bar{q} = \gamma(1)$. All these minimizing curves together constitute a piecewise-smooth curve, broken geodesic, connecting the point p_0 with the point $\bar{q}$. Let us denote this broken curve by u^z and its length by $J(z)$. We choose the parameter t on the curve u^z to be proportional to the length $s^z(t)$ of the arc of the curve u^z from the point p_0 to the point $u^z(t)$; specifically, we choose it so that

$$s^z(t) = \frac{J(z)}{1-\theta}(t-\theta).$$

With this choice of parameter,

$$u^z(\theta) = p_0, \quad u^z(1) = \bar{q}.$$

We denote by $t_i(z)$ the value of the parameter t corresponding to the point q_i. Thus,

$$u^z(t_i(z)) = q_i, \quad i = 1, \ldots, s.$$

We also introduce functions $t_0(z)$ and $t_{s+1}(z)$ and assume $t_0(z) = \theta$

and $t_{s+1}(z)=1$ for arbitrary $z \in U$. Obviously, for every $j = 0, 1, \ldots, s+1$, the functions $t_i(z)$ are smooth functions of the point $z \in U$. Since the parameter t can be expressed as a linear function of the arc length, each segment

$$u_i^z(t) = u^z(t)\big|_{[t_i(z),\, t_{i+1}(z)]}, \quad i=0,\ 1,\ \ldots,\ s,$$

of the broken curve u^z is a geodesic.

Let us now define, for an arbitrary point $z \in R^{s(m-1)}$ and arbitrary $i=0, 1, \ldots, s$, the variation φ_i^z of the geodesic $\gamma|_{[t_i, t_{i+1}]}$, by setting

$$\varphi_i^z(t,\ \tau) = u^{\tau z}(t), \quad |\tau| < \tau_0, \quad t_i(\tau z) \leqslant t \leqslant t_{i+1}(\tau z),$$

where τ_0 is a small positive number such that $\tau z \in U$ for $|\tau| < \tau_0$. (Such a τ_0 exists since the set U is open in the space $R^{s(m-1)}$ and contains the point 0.) Since the curve $u^{\tau z}(t)$, for $t_i(\tau z) \leqslant t \leqslant t_{i+1}(\tau z)$, is a geodesic, the variation φ_i^z is a Jacobian variation (more precisely, the restriction of some Jacobian variation to the set $|\tau| < \tau_0$, where $t_i(\tau z) \leqslant t \leqslant t_{i+1}(\tau z)$, of points $(t,\ \tau)$). Therefore, the vector field $X_i^z(t)$, for $t_i \leqslant t \leqslant t_{i+1}$, associated with it is a Jacobi field.

Define $t_i^z(\tau) = t_i(\tau z)$ and let

$$a_i^z(\tau) = u^{\tau z}\left(t_i^z(\tau)\right), \quad |\tau| < \tau_0,$$

be the left bounding curve of the variation φ_i^z. Then, as we know (cf. section 1),

$$\dot{a}_i^z(0) = \dot{t}_i^z(0)\,\dot{\gamma}_i + X_i^z(t_i), \tag{4}$$

where $\dot{\gamma}_i = \dot{\gamma}(t_i)$. But the same curve is (for $i > 0$) the right bounding curve of the variation φ_{i-1}^z, so that

$$\dot{a}_i^z(0) = \dot{t}_i^z(0)\,\dot{\gamma}_i + X_{i-1}^z(t_i).$$

Comparing these variations, we obtain the equation

$$X_{i-1}^z(t_i) = X_i^z(t_i),$$

for $i=1, \ldots, s$, which means that the fields $X_i^z(t)$ constitute on the interval $[\theta, 1]$ some broken Jacobi field $X^z(t)$. Obviously, this field will vanish at the points $t=\theta$ and $t=1$:

$$X^z(\theta) = 0, \quad X^z(1) = 0.$$

Furthermore, by virtue of our choice of the parameter t on the curves u^z, the function $k(\tau)$ corresponding to the variation φ_i^z (cf. section 2) is obviously given by the formula

$$k(\tau) = \frac{J^z(\tau)}{1-\theta}, \tag{5}$$

where

$$J^z(\tau) = J(\tau z).$$

In particular, this function is independent of t (and of i). Thus, the variation φ_i^z satisfies assumption (α) in section 2. Therefore, all the results of section 2 are valid for it. In particular,

$$\dot{k}(0) = \frac{1}{k(0)} \left(\frac{\nabla X_i^z}{dt}(t), \ \dot{\gamma}(t) \right)$$

[cf. section 2, formula (1)]. But, in accordance with formula (5),

$$\dot{k}(0) = \frac{\dot{J}^z(0)}{1-\theta},$$

so that

$$\left(\frac{\nabla X_i^z}{dt}(t), \ \dot{\gamma}(t) \right) = \frac{k(0)}{1-\theta} \dot{J}^z(0)$$

for arbitrary $i = 0, 1, \ldots, s$.

On the other hand,

$$J^z(\tau) = J_0^z(\tau) + \ldots + J_s^z(\tau),$$

where

$$J_i^z(\tau), \quad i = 0, 1, \ldots, s,$$

is the length of the curve $u^{\tau z}(t)$ for $t_i(\tau z) \leqslant t \leqslant t_{i+1}(\tau z)$. Since, in accordance with formula (4) of section 1,

$$\dot{J}_i^z(0) = \frac{(\dot{a}_{i+1}(0), \ \dot{\gamma}_{i+1}) - (\dot{a}_i(0), \ \dot{\gamma}_i)}{k(0)}, \tag{6}$$

it follows that

$$\dot{J}^z(0) = 0,$$

because $\dot{a}_0(0) = 0$ and $\dot{a}_{s+1}(0) = 0$. Thus,

$$\left(\frac{\nabla X_i^z}{dt}(t), \ \dot{\gamma}(t) \right) = 0 \text{ for every } i = 0, 1, \ldots, s \tag{7}$$

and, therefore,

$$\frac{d}{dt}\left(X_i^z(t), \ \dot{\gamma}(t) \right) = 0.$$

$\left(\text{Since } \frac{\nabla \dot{\gamma}}{dt}(t) = 0 \right)$; that is,

$$\left(X_i^z(t), \ \dot{\gamma}(t) \right) = a = \text{const.}$$

But, by virtue of equation (4), the constant a is independent of i and, consequently, equal to 0 since, for example, $X_0^z(0) = 0$. Thus, we have shown that

$$\left(X^z(t), \ \dot{\gamma}(t) \right) = 0 \text{ for every } t \in [\theta, \ 1],$$

that is, that

For an arbitrary point $z \in R^{s(m-1)}$, *the broken Jacobi field* $X^z(t)$ *belongs to the space* Θ.

9. A theorem on isomorphism

The construction described in section 8 defines a mapping

$$R^{s(m-1)} \to \Theta, \tag{1}$$

that assigns to every point $z \in R^{s(m-1)}$ the field $X^z(t) \in \Theta$.

We have

The mapping (1) is an isomorphism of the space $R^{s(m-1)}$ *onto the space* Θ.

To prove this proposition, it will be sufficient for us to prove the following three assertions:

(1) the mapping (1) is linear;

(2) the mapping (1) is monomorphic; that is, $X^z(t) = 0$ only when $z = 0$;

(3) the dimension of the space Θ does not exceed the dimension $s(m-1)$ of the space $R^{s(m-1)}$.

With this in view, let us consider the direct sum

$$L = M_{p_1} + \ldots + M_{p_s}$$

of tangent spaces $M_{p_1}, \ldots, M_{p_s}$ of the manifold M at the points $p_1 = \gamma(t_1), \ldots, p_s = \gamma(t_s)$, respectively, and let us consider the natural mapping

$$\Theta \to L, \tag{2}$$

which assigns to every field $X(t) \in \Theta$ the vector

$$(X(t_1), \ldots, X(t_s))$$

in the space L. Obviously, this mapping is linear. Furthermore, it is monomorphic because the points t_i and t_{i+1} are nonconjugate and hence the Jacobi field

$$X_i(t) = X(t)|_{[t_i, t_{i+1}]}$$

is unambiguously defined by the vectors $X(t_i)$ and $X(t_{i+1})$. (In the case $i = 0$, it is determined only by the vector $X(t_1)$ because, by hypothesis, $X(0) = 0$; in the case $i = s$, it is determined only by the vector $X(t_s)$ because, by hypothesis, $X(1) = 0$).

On the other hand, we can construct a mapping

$$R^{s(m-1)} \to L, \tag{3}$$

that assigns to every point $z \in R^{s(m-1)}$ the tangent vectors

$$\dot{a}_1^z(0), \ldots, \dot{a}_s^z(0)$$

of the bounding curves $a_i^z(t)$ of the variations φ_i^z, for $i=1, \ldots, s$. (We note that these vectors are tangent to the corresponding submanifolds $P_1, \ldots, P_s$ because $a_i^z(\tau) \in P_i$ for arbitrary τ and arbitrary $i=1, \ldots, s$.) Let e_k^j, for $j=1, \ldots, s$ and $k=1, \ldots, m-1$, denote unit vectors in the space $R^{s(m-1)}$, that is, vectors all the coordinates of which are equal to 0 with the exception of the coordinate z_j^k which is equal to 1. Let

$$(A_{1k}^j, \ldots, A_{sk}^j), \; j=1, \ldots, s; \; k=1, \ldots, m-1,$$

denote vectors in the space L corresponding, under the mapping (3), to the vectors e_k^j. Since $a_i^z(\tau)=f_i(\tau z)$, where $f_i(z)=u^z(t_i(z))$, it follows from the familiar formal rules for differentiating functions of a vector argument that, for arbitrary $z=z_j^k e_k^j$,

$$\dot{a}_i^z(0)=z_j^k A_{ik}^j.$$

(We recall that the vector A_{ik}^j in the space M_{p_i} is the vector $\dot{a}_i^z(0)$ corresponding to the vector $z=e_k^j$.) Consequently, the mapping (3) is linear. Since, for arbitrary $i=1, \ldots, s$, the vectors $A_{i1}^i, \ldots, A_{im-1}^i$ (where we do not follow the summation convention) are linearly independent (obviously, they constitute a basis for the tangent space $(P_i)_{p_i}$), the mapping (3) is monomorphic.

As we know, for arbitrary $z \in R^{s(m-1)}$ and arbitrary $i=1, \ldots, s$,

$$X^z(t_i)=\dot{a}_i^z(0)-\dot{t}_i^z(0)\,\dot{\gamma}_i$$

(cf. formula (4) of section 8). If we take the scalar product of both sides of the equation by $\dot{\gamma}_i$ and take into account the orthogonality of the vectors $X^z(t_i)$ and $\dot{\gamma}_i$, we obtain

$$\dot{t}_i^z(0)=\frac{(\dot{a}_i^z(0), \dot{\gamma}_i)}{(\dot{\gamma}_i, \dot{\gamma}_i)}.$$

Consequently,

$$X^z(t_i)=\dot{a}_i^z(0)-\frac{(\dot{a}_i^z(0), \dot{\gamma}_i)}{(\dot{\gamma}_i, \dot{\gamma}_i)}\dot{\gamma}_i.$$

This means that the mapping

$$L \to L, \tag{4}$$

which assigns to each vector $(A_1, \ldots, A_s) \in L$ the vector $(B_1, \ldots, B_s) \in L$ defined by

$$B_i=A_i-\frac{(A_i, \dot{\gamma}_i)}{(\dot{\gamma}_i, \dot{\gamma}_i)}\dot{\gamma}_i,$$

has the property that the composite of this mapping and the mapping (3) coincides with the composite of the mappings (1) and (2):

$$\begin{array}{ccc} R^{s(m-1)} & \xrightarrow{(1)} & \Theta \\ \downarrow{\scriptstyle (3)} & & \downarrow{\scriptstyle (2)} \\ L & \xrightarrow{(4)} & L \end{array}$$

In other words, if we consider the monomorphisms (2) and (3) as embeddings, the mapping (1) will be the restriction of the mapping (4). Since the mapping (4) is obviously linear, its restriction (1) is also linear [cf. assertion (1)].

The kernel of the mapping (4) consists of the vectors $(A_1, \ldots, A_s) \in L$ such that, for arbitrary $i = 1, \ldots, s$, the vector A_i is proportional to the vector $\dot{\gamma}_i$. On the other hand, for arbitrary $i = 1, \ldots, s$, the vector $\dot{a}_i^z(0)$, being a tangent vector of the submanifold P_i, is proportional to the vector $\dot{\gamma}_i$ if and only if it is equal to 0. Consequently, the kernel of the mapping (4) intersects the image of the mapping (3) only at 0; that is, the composite of the mappings (3) and (4) is a monomorphism. Therefore, the mapping (1) is also a monomorphism [cf. assertion (2)].

Finally, since $(X(t), \dot{\gamma}(t)) = 0$ for an arbitrary field $X(t) \in \Theta$, the mapping (2) maps the space Θ into the subspace L_0 of the space L consisting of the vectors $(A_1, \ldots, A_s) \in L$ such that

$$(A_i, \dot{\gamma}_i) = 0 \quad \text{for every } i = 1, \ldots, s. \tag{5}$$

Since the subspace L_0 is obviously $s(m-1)$-dimensional and the mapping (2) is monomorphic, the dimension of the space does not exceed $s(m-1)$ [assertion (3)]. This completes the proof of the theorem given above.

Remark: Since the dimensions of the spaces Θ and L_0 coincide, the monomorphism

$$\Theta \to L_0$$

is an isomorphism; that is,

For an arbitrary set of vectors

$$A_1 \in M_{p_1}, \ \ldots, \ A_s \in M_{p_s},$$

that satisfy condition (5), there exists exactly one field $X(t) \in \Theta$ *such that*

$$X(t_i) = A_i$$

for all $i = 1, \ldots, s$.

10. Morse's quadratic form

Keeping the notations used in the last two sections, let us consider the second derivatives

$$\left(\frac{\partial^2 J(z)}{\partial z_{i_1}^{k_1} \partial z_{i_2}^{k_2}} \right)_{z=0}, \quad \begin{array}{l} i_1, \ i_2 = 1, \ \ldots, \ s, \\ k_1, \ k_2 = 1, \ \ldots, \ m-1, \end{array}$$

of the function $J(z)$ at the point $z = 0$. (Since $\dot{J}^z(0) = \frac{\partial J(0)}{\partial z_i^k} z_i^k$ for all z, the first derivatives of the function $J(z)$ vanish at $z = 0$.) If we set

$$Q_\theta(z) = \frac{1}{l} I_\theta^1(X^z(t))$$

we define a quadratic form $Q_\theta(z)$ [the so-called *Hessian* of the function $J(z)$] on the space $R^{s(m-1)}$ We shall call this form *Morse's quadratic form*. It depends on the choice of points (1), the submanifolds (2), and the local coordinates (3) of section 8.

Since

$$\frac{d^2J^z(\tau)}{d\tau^2} = \frac{\partial^2 J(\tau z)}{\partial z_{i_1}^{k_1} \partial z_{i_2}^{k_2}} z_{i_1}^{k_1} z_{i_2}^{k_2},$$

for the function $J^z(\tau) = J(\tau z)$, we have

$$Q_\theta(z) = \ddot{J}^z(0),$$

and, hence,

$$Q_\theta(z) = \ddot{J}_0^z(0) + \ldots + \ddot{J}_s(0),$$

where, as in section 8, $J_i^z(\tau)$ is the length of the curve $u^{\tau z}(t)$ for $t_i(\tau z) \leqslant t \leqslant t_{i+1}(\tau z)$, with $i = 0, 1, \ldots, s$.

On the other hand, the number $\ddot{J}_i^z(0)$ is obviously the second variation of the arc length of the geodesic $\gamma|_{[t_i, t_{i+1}]}$ corresponding to the variation φ_i^z. Since, as we know, the variation φ_i^z satisfies assumption (α) of section 2, formula (2) of section 2 is valid for the number $\ddot{J}_i^z(0)$. Here, by virtue of equation (7) of section 8, the integrand in this formula is simplified and the integral is transformed (up to a constant factor $1/l$, where $l = k(0)$ is the length of the segment $\gamma|_{[0,1]}$ of the geodesic γ) into the functional $I_{t_i}^{t_{i+1}}$. Thus,

$$\ddot{J}_i^z(0) = \frac{1}{l}\left[\left(\frac{\nabla \dot{a}_{i+1}}{d\tau}(0), \dot{\gamma}_{i+1}\right) - \left(\frac{\nabla \dot{a}_i}{d\tau}(0), \dot{\gamma}_i\right)\right] + \frac{1}{l} I_{t_i}^{t_{i+1}}(X_i^z(t)).$$

Summing over i and noting that

$$\frac{\nabla \dot{a}_0}{d\tau}(0) = 0, \quad \frac{\nabla \dot{a}_{s+1}}{d\tau}(0) = 0,$$

we immediately obtain from the above equation

$$Q_\theta(z) = \frac{1}{l} I_\theta^1(X^z(t)). \tag{1}$$

In accordance with the theorem of section 9, the field $X^z(t)$ is nonzero for $z \neq 0$. Therefore, in accordance with inequality (2) of section 7 (this inequality is applicable since $X^z(\theta) = 0$ and $X^z(1) = 0$), it follows from formula (1) that

For θ sufficiently close to unity, the form Q_θ is positive-definite.

In what follows, we shall be interested in two numerical characteristics of the form $Q_\theta(z)$, namely, its *corank* $c_\theta = c(Q_\theta)$, that is, the dimension of the maximum subspace $C_\theta \subset R^{s(m-1)}$ on which the form $Q_\theta(z)$ is idenfically equal to 0, and its *negative index of inertia* $h_\theta = h(Q_\theta)$, that is, the dimension of the maximum subspace $H_\theta \subset R^{s(m-1)}$ on which the form $Q_\theta(z)$ is negative-definite.

It turns out that

The numbers c_θ and h_θ are independent of the arbitrariness attending the construction of the form $Q_\theta(z)$ and are determined exclusively by the number θ (and the geodesic γ).

In other words, no matter how we may choose the numbers t_i, the submanifolds P_i, and the local coordinates z_i^k (while, of course, we keep all the necessary conditions and, in particular, the condition that the points $t_1, \ldots, t_s$ be sufficiently dense on the interval $[\theta, 1]$), the numbers c_θ and h_θ for the corresponding forms $Q_\theta(z)$ will have the same values.

Obviously, it will be sufficient to prove this assertion in the following special form:

Suppose that we add to the points t_i defining the form $Q_\theta(z)$ one more point, (let us call it t^*) in the interval (t_{i_0}, t_{i_0+1}) and let us construct for the thus augmented system of points the corresponding form $Q_\theta^*(z)$ [by choosing the submanifold P^* and the local coordinates z_*^k on that submanifold arbitrarily at the point $\gamma(t^*)$]. Then, the numbers $c_\theta^* = c(Q_\theta^*)$ and $h_\theta^* = h(Q_\theta^*)$ will be equal to the numbers $c_\theta = c(Q_\theta)$ and $h_\theta = h(Q_\theta)$, respectively.

Let us first prove this assertion for the number h_θ. Just as above, let H_θ denote the maximum subspace of the space $R^{s(m-1)}$ on which the form $Q_\theta(z)$ is negative-definite. Analogously, let H_θ^* denote the maximum subspace of the space $R^{(s+1)(m-1)}$ on which the form $Q_\theta^*(z)$ is negative-definite. Furthermore, let Θ denote the space (considered in sections 8 and 9) of broken Jacobi fields and let Θ^* denote the analogous space corresponding to the augmented system of points t_i. Obviously,

$$\Theta \subset \Theta^* \tag{2}$$

Therefore, the isomorphism $R^{s(m-1)} \to \Theta$, which we considered in section 9, and the analogous isomorphism $R^{(s+1)(m-1)} \to \Theta^*$ define a monomorphism

$$R^{s(m-1)} \to R^{(s+1)(m-1)}. \tag{3}$$

It follows immediately from formula (1) that, if a point z in the space $R^{s(m-1)}$ is mapped under the monomorphism (3) into a point z^* of the space $R^{(s+1)(m-1)}$, then

$$Q_\theta(z) = Q_\theta^*(z^*)$$

(since each of these numbers is equal to $I_\theta^1(X^z(t))1/t)$. Therefore, the monomorphism (3) maps the subspace H_θ into the subspace H_θ^*.

Now, consider the natural projection

$$p : R^{(s+1)(m-1)} \to R^{s(m-1)}, \tag{4}$$

under which we discard the coordinates z_*^i corresponding to the added point t^*. Suppose that $z^* \in R^{(s+1)(m-1)}$ and suppose that the point $z \in R^{s(m-1)}$ is the image of the point z^* under the projection (4). The broken Jacobi fields $X^z(t)$ and $X^{z^*}(t)$ corresponding to the points z and z^* coincide everywhere on the interval $[\theta, 1]$ except possibly in the interval (t_{i_0}, t_{i_0+1}). On this interval, the field $X^z(t)$ is a smooth Jacobi field but the field $X^{z^*}(t)$ may have a break at the point t^*. Therefore, on the basis of the minimal property of Jacobi fields (cf. section 7),

$$I_{t_{i_0}}^{t_{i_0+1}}(X^z(t)) \leqslant I_{t_{i_0}}^{t_{i_0+1}}(X^{z^*}(t)),$$

and, therefore, in accordance with formula (1),

$$Q_\theta(z) \leqslant Q_\theta^*(z^*).$$

This means, in particular, that the projection (4) maps the subspace H_θ^* into the space H_θ.

Let $z^* \in H_\theta^*$ denote any nonzero point in the kernel of the projection (4) that belongs to the subspace H_θ^*. The corresponding field $X^{z^*}(t)$ vanishes everywhere on the interval $[\theta, 1]$ except in the interval (t_{i_0}, t_{i_0+1}). Moreover, since $z^* \in H_\theta^*$, we have

$$I_{t_{i_0}}^{t_{i_0+1}}(X_{i_0}^{z^*}(t)) = I_\theta^1(X^{z^*}(t)) = lQ_\theta^*(z^*) < 0.$$

But this is impossible since, in accordance with inequality (2) of section 7, we have

$$I_{t_{i_0}}^{t_{i_0+1}}(X_{i_0}^{z^*}(t)) > 0.$$

This contradiction proves that the kernel of the projection (4) intersects the subspace H_θ^* only at 0, that is, that the projection (4) is monomorphic on H_θ^*.

Since the composite of the monomorphism (3) and the projection (4) is obviously the identity mapping of the space $R^{s(m-1)}$, we have shown that the mappings (3) and (4) define inverse isomorphisms

$$H_\theta \leftrightarrows H_\theta^*.$$

Consequently,

$$h_\theta = h_\theta^*.$$

The proof is word for word the same for the number c_θ.

11. Evaluation of the index of a point with the aid of Morse's form

The relationship between Morse's form and the concept of a conjugate point is brought out by the following proposition:

A point $t=\theta$ is conjugate to the point $t = 1$ if and only if the form Q_θ is degenerate.

The "degree of conjugacy" of the point $t=\theta$ is measured by its index $\lambda_1(\theta)$, and the "degree of degeneracy" of the form Q_θ by its corank c_θ. It turns out that

The index $\lambda_1(\theta)$ is equal to the corank c_θ.

To prove these assertions, let us consider the subspace Θ_0 of the space Θ consisting of smooth Jacobi fields. Obviously,

$$\Theta_0 = \Lambda_\gamma(\theta) \cap \Lambda_\gamma(1)$$

(because, on the one hand, all the fields $X(t) \in \Theta_0$ are, by hypothesis, equal to 0 both at $t=\theta$ and at $t=1$ and, on the other hand, an arbitrary field $X(t) \in \Lambda_\gamma(\theta) \cap \Lambda_\gamma(1)$ has the property that $(X(t), \gamma(t)) = 0$ [cf. formula (6) of section 3]. Therefore, the index $\lambda_1(\theta)$ is equal to the dimension of the subspace Θ_0).

At the same time, the corank c_θ is, by definition, equal to the dimension of the maximum subspace $C_\theta \subset R^{s(m-1)}$ on which the form Q_θ vanishes identically. Therefore, to prove the assertions made, it will be sufficient to show that

The isomorphism $R^{s(m-1)} \to \Theta$ referred to in section 9 maps the subspace C_0 onto the subspace Θ_0; in other words, the field $X^z(t)$, where $z \in R^{s(m-1)}$, belongs to the subspace Θ_0 if and only if $z \in C_\theta$.

Keeping this in mind, let us transform the fundamental formula (1) of section 10 for the form $Q_\theta(z)$ with the aid of formula (3) of section 7. Since, by virtue of this last formula

$$I_{t_i}^{t_{i+1}}(X_i^z(t)) = \left(\frac{\nabla X_i^z}{dt}(t_{i+1}), X_i^z(t_{i+1})\right) - \left(\frac{\nabla X_i^z}{dt}(t_i), X_i^z(t_i)\right) = \\ = \left(\frac{\nabla X^z}{dt}(t_{i+1}-0), X^z(t_{i+1})\right) - \left(\frac{\nabla X^z}{dt}(t_i+0), X^z(t_i)\right),$$

we have

$$Q_\upsilon(z) = \frac{1}{l}\sum_{i=1}^{s}\left(\frac{\nabla X^z}{dt}(t_i-0) - \frac{\nabla X^z}{dt}(t_i+0), X^z(t_i)\right),$$

because

$$X^z(t_0) = X^z(\theta) = 0, \text{ and } X^z(t_{s+1}) = X^z(1) = 0.$$

Let $X_k^j(t)$ denote the basis of the space Θ corresponding under the isomorphism (1) of section 9 to the standard basis e_k^j of the space $R^{s(m-1)}$, where $j=1, \ldots, s$ and $k=1, \ldots, m-1$. Then, for an arbitrary point $z \in R^{s(m-1)}$, we have

$$X^z(t) = z_j^k X_k^j(t).$$

If we differentiate this equation, we obtain

$$\frac{\nabla X^z}{dt}(t_i \pm 0) = z^k_j \frac{\nabla X^j_k}{dt}(t_i \pm 0), \qquad i = 1, \ldots, s.$$

Therefore,

$$Q_\theta(z) = \frac{1}{l}\left[\sum_{i=1}^{s}\left(\frac{\nabla X^{j_1}_{k_1}}{dt}(t_i - 0) - \frac{\nabla X^{j_1}_{k_1}}{dt}(t_i + 0),\ X^j_k(t_i)\right)\right] z^{k_1}_{j_1} z^k_j.$$

From this, we obtain for the partial derivatives

$$L^j_k(z) = \frac{\partial Q_\theta(z)}{\partial z^k_j}$$

the following expression:

$$L^j_k(z) = \frac{2}{l}\left[\sum_{i=1}^{s}\left(\frac{\nabla X^{j_1}_{k_1}}{dt}(t_i - 0) - \frac{\nabla X^{j_1}_{k_1}}{dt}(t_i + 0),\ X^j_k(t_i)\right)\right] z^{k_1}_{j_1} =$$

$$= \frac{2}{l}\sum_{i=1}^{s}\left(\frac{\nabla X^z}{dt}(t_i - 0) - \frac{\nabla X^z}{dt}(t_i + 0),\ X^j_k(t_i)\right).$$

We know from elementary linear algebra that the point z belongs to the subspace C_θ if and only if

$$L^j_k(z) = 0 \qquad \text{for all} \quad j \text{ and } k.$$

Consequently, if $z \in C_\theta$, we have

$$\sum_{i=1}^{s}\left(\frac{\nabla X^z}{dt}(t_i - 0) - \frac{\nabla X^z}{dt}(t_i + 0),\ X^j_k(t_i)\right) = 0.$$

But, for arbitrary $i = 1, \ldots, s$ and arbitrary $k = 1, \ldots, m-1$, the vectors $X^j_k(t_i)$, for $j \neq i$, are obviously equal to 0, so that

$$\left(\frac{\nabla X^z}{dt}(t_i - 0) - \frac{\nabla X^z}{dt}(t_i + 0),\ X^i_k(t_i)\right) = 0. \tag{1}$$

(Here, we are *not* summing over i.) On the other hand, for arbitrary $i = 1, \ldots, s$, the vectors $X^i_k(t_i)$, for $k = 1, \ldots, m-1$ (again we are not summing over i) together with the vector $\dot{\gamma}_i = \dot{\gamma}(t_i)$ constitute, as one can easily see, a basis for the space M_{p_i}. Furthermore, in accordance with formula (6) of section 8,

$$\left(\frac{\nabla X^z}{dt}(t_i \pm 0),\ \dot{\gamma}_i\right) = 0 \qquad \text{for all} \qquad i = 1, \ldots, s.$$

Therefore, equation (1) is possible only when

$$\frac{\nabla X^z}{dt}(t_i - 0) = \frac{\nabla X^z}{dt}(t_i + 0), \qquad i = 1, \ldots, s,$$

that is, only when the field $X^z(t)$ is differentiable. But it is easy to see that

An arbitrary differentiable broken Jacobi field is smooth.

This is true because the one-sided second covariant derivatives

$$\frac{\nabla}{dt}\frac{\nabla}{dt}X(t_i \pm 0)$$

of an arbitrary Jacobi field $X(t)$ coincide, on the basis of equation (3) of section 3 (applied respectively on the intervals $[t_{i-1}, t_i]$ and $[t_i, t_{i+1}]$), with the same vector $-R_{\gamma(t_i)}(X(t_i), \dot{\gamma}_i)\dot{\gamma}_i$ and hence are equal. Consequently, if the field $X(t)$ is differentiable, it is twice differentiable. Furthermore, since the field $X(t)$ satisfies equation (3) of section 3 everywhere, it is a smooth field (cf. section 5).

Thus, the field $X^z(t)$ is smooth, that is, belongs to the subspace Θ_0.

Conversely, if $\dot{X}^z(t) \in \Theta_0$, then

$$\frac{\nabla X^z(t)}{dt}(t_i - 0) = \frac{\nabla X^z}{dt}(t_i + 0) \qquad \text{for every } i = 1, \ldots, s$$

and, therefore, $L_k^j(z) = 0$ for all j and k: that is, $z \in C_\theta$. This completes the proof of the assertion made at the beginning of this section.

12. Evaluation of the index of an interval with the aid of Morse's form

We shall denote the form Q_θ evaluated for $\theta = 0$ by the letter Q. According to the assertion proven in section 11,

The index $\lambda_1(0)$ of the point t = 0 (with respect to the point t = 1) is equal to the corank $c = c_0$ of the form $Q = Q_0$.

The principal purpose of the present section consists in proving the following proposition (known as "Morse's index theorem"):

The index $\lambda_1(0, 1)$ of an open interval (0, 1) (with respect to the point t = 1) is equal to the negative index of inertia $h = h_0$ of the form Q.

This theorem, used in conjunction with the preceding assertion, enables us to evaluate the index of an arbitrary (open or closed) interval of the real axis since every such interval can be represented as the difference of two intervals with common end-point.

Let us first prove the inequality

$$h \leqslant \lambda_1(0, 1). \tag{1}$$

With this in mind, let us return to the forms $Q_\theta(z)$ with arbitrary $\theta \in [0, 1)$. By construction, these forms depend on the choice of the numbers

$$\theta < t_1 < \ldots < t_s < 1,$$

on the manifolds

$$P_1, \ldots, P_s,$$

passing through the points $p_1 = \gamma(t_1), \ldots, p_s = \gamma(t_s)$, and on the systems of local coordinates

$$z_i^1, \ldots, z_i^{m-1}, \quad i = 1, \ldots, s,$$

on these submanifolds. Now, we shall specialize our construction by choosing these given conditions "linear in θ" for all θ.

Specifically, we assume that these objects are chosen for $\theta = 0$. Suppose that they are the numbers

$$0 < \bar{t}_1 < \ldots < \bar{t}_s < 1,$$

the submanifolds

$$\bar{P}_1, \ldots, \bar{P}_s$$

and the local coordinates

$$\bar{z}_i^1, \ldots, \bar{z}_i^{m-1}, \quad i = 1, \ldots, s.$$

Then, for arbitrary $\theta = [0, 1)$, we set

$$t_i = \theta + \bar{t}_i - \theta\bar{t}_i, \quad i = 1, \ldots, s.$$

Obviously, the numbers $\bar{t}_1, \ldots, \bar{t}_s$ can be chosen sufficiently thickly (in the sense of section 8) on the interval [0, 1] that, for an arbitrary number θ, the numbers $t_1, \ldots, t_s$ will also be sufficiently thickly distributed (though on the interval $[\theta, 1]$).

Let

$$\tau_i \colon M_{\bar{p}_i} \to M_{p_i}$$

be a parallel translation along the geodesic γ from the point $\bar{p}_i = \gamma(\bar{t}_i)$ to the point $p_i = \gamma(t_i)$. For every $i = 1, \ldots, s$, let us choose the submanifold P_i in such a way that its tangent space will be obtained by a translation τ_i from the tangent space of the submanifold $\bar{P}_i$. Obviously, such a choice of submanifolds P_i is always possible (while all the other conditions are preserved).

Finally, we choose the local coordinates

$$z_i^1, \ldots, z_i^{m-1}$$

on the submanifold P_i so that the vector fields

$$\frac{\partial}{\partial z_i^1}, \ldots, \frac{\partial}{\partial z_i^{m-1}}$$

corresponding to them are obtained from the fields

$$\frac{\partial}{\partial \bar{z}_i^1}, \ldots, \frac{\partial}{\partial \bar{z}_i^{m-1}}:$$

by a translation τ_i:

$$\frac{\partial}{\partial z_i^1} = \tau_i\left(\frac{\partial}{\partial \bar{z}_i^1}\right), \ \ldots, \ \frac{\partial}{\partial z_i^{m-1}} = \tau_i\left(\frac{\partial}{\partial \bar{z}_i^{m-1}}\right).$$

With this choice of given conditions necessary for the construction of the form $Q_\theta(z)$, it is obvious that

The coefficients of the form Q_θ are continuous functions of the parameter θ.

Therefore, as the parameter θ varies continuously, the negative index of inertia h_θ of the form Q_θ can vary only when the parameter passes through that value θ_0 at which the form Q_θ is degenerate. Then, the change in the index cannot exceed the corank of the form Q_{θ_0}. By virtue of the results of the preceding section, it then follows immediately that the total variation in the index of the form Q_θ as the parameter varies continuously from some value θ_0 (close to one) to 0 does not exceed the index $\lambda_1(0, 1)$ of the interval $(0, 1)$. On the other hand, we know that when θ_0 is close to one, the form $Q_\theta(z)$ is positive-definite (cf. section 10) and hence $h_{\theta_0} = 0$. Consequently, the variation in the index h_θ as θ varies from θ_0 to 0 continuously is equal to the index $h = h_0$ of the form $Q = Q_0$. This completes the proof of inequality (1).

Thus, it remains for us to prove only the opposite inequality

$$h \geqslant \lambda_1(0, 1). \tag{2}$$

Since this inequality is automatically satisfied for $\lambda_1(0, 1) = 0$, we may assume that $\lambda_1(0, 1) > 0$, that is, that the interval $(0, 1)$ contains points conjugate to the point $t = 1$. Suppose that

$$0 < a_1 < \ldots < a_r < 1$$

represent all these points and that

$$\lambda_1, \ \ldots, \ \lambda_r$$

are their indices. Suppose, furthermore, that

$$A_{j,1}(t), \ \ldots, \ A_{j,\lambda_j}(t), \qquad j = 1, \ \ldots, \ r$$

is an arbitrary basis in the space $\Lambda_\gamma(a_j) \cap \Lambda_\gamma(1)$. As we know (cf. formula (6) of section 3), for all t and arbitrary j and k, we have

$$(A_{j,k}(t), \dot{\gamma}(t)) = 0. \tag{3}$$

Let us suppose that the following additional conditions are imposed on the choice of the points t_i (cf. section 8) that enter into the construction of the form Q:

(1) for arbitrary $j = 0, \ldots, r$ the interval $(a_j a_{j+1})$ [where we take $a_0 = 0$ and $a_{r+1} = 1$] contains at least one point t_i;

(2) no point t_i, for $i = 1, \ldots, s$, coincides with any point a_j, for $j = 1, \ldots, r$. Since the index h is independent of the choice of the

points $t_1, \ldots, t_s$ (as shown in section 10), these supplementary conditions have no effect on the generality of the result.

Accepting these conditions, let us look at the piecewise-smooth field $X_{j,k}(t)$, for $j=1, \ldots, r$ and $k=1, \ldots, \lambda_j$, defined on the interval $[0, 1]$ by

$$X_{j,k}(t) = \begin{cases} 0 & \text{if} \quad 0 \leqslant t \leqslant a_j, \\ A_{j,k}(t) & \text{if} \quad a_j \leqslant t \leqslant 1. \end{cases}$$

Since, by virtue of formula (3),

$$(X_{j,k}(t_i), \dot{\gamma}_i) = 0 \quad \text{for every} \quad i = 1, \ldots, s,$$

there exists, by virtue of the remark made at the end of section 9, one and only one broken Jacobi field

$$X_{j,k}(t) \in \Theta, \quad 0 = \theta \leqslant t \leqslant 1,$$

such that

$$\overline{X}_{j,k}(t_i) = X_{j,k}(t_i).$$

Let $z_{j,k}$ denote that (unique) point in the space $R^{s(m-1)}$ such that

$$\overline{X}_{j,k}(t) = X^{z_{j,k}}(t).$$

Let us show that

The points $z_{j,k}$, for $j=1, \ldots, r$ and $k=1, \ldots, \lambda_j$ are linearly independent.

Proof: Let $a^{j,k}$ be real numbers such that

$$a^{j,k} z_{j,k} = 0.$$

Then, the field $a^{j,k}\overline{X}_{j,k}(t)$ is identically equal to 0 and, hence, the field

$$X(t) = a^{j,k} X_{j,k}(t)$$

has the property that

$$X(t_i) = 0 \quad \text{for every} \quad i = 1, \ldots, s.$$

Let l denote the smallest i such that $t_l > a_1$ [cf. condition (1)]. Obviously, the field $X(t)$ is equal on the segment $[a_1, t_l]$ to the field $a^{1,k}A_{1,k}(t)$ and, therefore, it is a smooth Jacobi field and it is equal to 0 at $t=a_1$. Since the points a_1 and t_l are nonconjugate (because $a_1 \in [t_{l-1}, t_l]$) and the field $X(t)$ is equal to 0 at $t=t_l$, it follows that the field $X(t)$ is identically equal to 0 on the interval $[a_1, t_l]$. Therefore,

$$a^{1,k} = 0 \quad \text{for all} \quad k = 1, \ldots, \lambda_1.$$

If we repeat this reasoning for a_2, we obtain in an analogous manner

$$a^{2,k}=0 \quad \text{for all} \quad k=1, \ldots, \lambda_2.$$

After r such steps, we finally see that $a^{j,k}=0$ for all $j=1, \ldots, r$ and $k=1, \ldots, \lambda_j$. Consequently, the points $z_{j,k}$ are linearly independent.

Now, let Z denote the subspace generated by the points $z_{j,k}$ in the space $R^{s(m-1)}$. According to what we have proved, the dimension of this subspace is equal to the total number of points $z_{j,k}$, that is, equal to

$$\lambda_1+\ldots+\lambda_r=\lambda_1(0, 1).$$

Therefore, to prove inequality (2), it suffices to show that

The form Q is negative-definite on the subspace Z.

Let $z=a^{j,k}z_{j,k}$ denote an arbitrary point in the subspace Z. Consider the field

$$X(t)=a^{j,k}X_{j,k}(t).$$

This field is a smooth Jacobi field on each interval $[a_i, a_{i+1}]$ for $i=0, \ldots, r$ (just as above, $a_0=0$ and $a_{r+1}=1$). Therefore (cf. formula (3) of section 7),

$$I_{a_i}^{a_{i+1}}(X(t))=\left(\frac{\nabla X}{dt}(a_{i+1}-0), X(a_{i+1})\right)-\left(\frac{\nabla X}{dt}(a_i+0), X(a_i)\right).$$

Consequently,

$$I_0^1(X(t))=\sum_{i=1}^{r}\left(\frac{\nabla X}{dt}(a_i-0)-\frac{\nabla X}{dt}(a_i+0), X(a_i)\right),$$

because $X(a_0)=X(0)=0$ and $X(a_{r+1})=X(1)=0$.

Consider the expression

$$\left(\frac{\nabla X}{dt}(a_i-0)-\frac{\nabla X}{dt}(a_i+0), X(a_i)\right), \qquad i=1, \ldots, r. \tag{4}$$

Since the field $X(t)$ is a linear combination of the fields $A_{j,k}(t)$, where $j<i$, on the interval $[a_{i-1}, a_i]$ and is a linear combination of the fields $A_{j,k}(t)$, where $j \leqslant i$ (with the same coefficients of the fields $A_{j,k}(t)$, where $j<i$), on the interval $[a_i, a_{i+1}]$, the vector

$$\frac{\nabla X}{dt}(a_i-0)-\frac{\nabla X}{dt}(a_i+0)$$

is a linear combination of the vectors

$$\frac{\nabla A_{i,l}}{dt}(a_i), \quad l=1, \ldots, \lambda_i.$$

On the other hand, since $A_{i,k}(a_i)=0$, the vector $X(a_i)$ for $i>1$, is a linear combination of the vectors

$$A_{j,k}(a_i), \quad j=1, \ldots, i-1; \; k=1, \ldots, \lambda_j,$$

and is equal to 0 for $i = 1$. Consequently, the expression (4) is equal to 0 for $i=1$ and is a linear combination of expressions of the form

$$\left(\frac{\nabla A_{i,l}}{dt}(a_i), \; A_{j,k}(a_i)\right), \quad \begin{array}{l} l=1, \ldots, \lambda_i, \\ j=1, \ldots, i-1, \\ k=1, \ldots, \lambda_j \end{array}$$

for $i>1$. But, by hypothesis, all fields of the form $A_{j,k}$ belong to the space $\Lambda_\gamma(1)$ and, hence, in accordance with formula (5) of section 3,

$$\left(\frac{\nabla A_{i,l}}{dt}(a_i), \; A_{j,k}(a_i)\right) = \left(A_{i,l}(a_i), \; \frac{\nabla A_{j,k}}{dt}(a_i)\right) = 0,$$

because $A_{i,l}(a_i)=0$. Thus, the expressions (4) are equal to 0 for arbitrary i. Consequently,

$$I_0^1(X(t))=0.$$

On the other hand, since the field $X^z(t) \in \Theta$ corresponding to the point z is a smooth Jacobi field on every interval $[t_i, t_{i+1}]$, for $i=0, 1, \ldots, s$, and since $X^z(t_i)=X(t_i)$ and $X^z(t_{i+1})=X(t_{i+1})$, it follows that, in accordance with Theorem 7 (the conditions of applicability of which are, by hypothesis, satisfied for arbitrary $i=0, 1, \ldots, s$), we have the inequality

$$I_{t_i}^{t_{i+1}}(X^z(t)) \leqslant I_{t_i}^{t_{i+1}}(X(t)), \tag{5}$$

with equality holding only when $X(t)=X^z(t)$ on $[t_i, t_{i+1}]$. But, for $z \neq 0$, the equation $X(t)=X^z(t)$ on the interval $[t_i, t_{i+1}]$ for all $i=0, 1, \ldots, s$ is impossible since the field $X(t)$ has a break at least at one point a_j, whereas the field $X^z(t)$ is smooth at that point. (We recall that, by hypothesis, the points a_j are distinct from the points t_i). Therefore, if we sum inequalities (5) over i, we obtain

$$I_0^1(X^z(t)) < I_0^1(X(t)),$$

that is, in accordance with what was proven above,

$$I_0^1(X^z(t)) < 0.$$

Consequently, in accordance with formula (1) of section 10,

$$Q(z) < 0, \quad z \neq 0.$$

This completes the proof of the assertion made and, with it, Morse's theorem.

13. Bott's quadratic form. The final formulation of the theorems on indices

The construction of Morse's form $Q(z)$ includes a considerable element of arbitrariness consisting, in particular, in the choice of the submanifolds $P_1, \ldots, P_s$. Recently, Bott suggested consideration of a different quadratic form with the same end in view. The construction of Bott's form contains considerably less arbitrariness.

Analogously to the case of Morse's form, the construction of Bott's form begins with the choice of a system of points

$$0 = t_0 < t_1 < \ldots < t_s < t_{s+1} = 1$$

in the interval [0, 1]. These points, just as in the case of Morse's form, are chosen on the interval [0, 1] rather thickly, so thickly that, for each $i = 0, 1, \ldots, s$, the interval $[t_i, t_{i+1}]$ will contain no points conjugate with the point t_{i+1}. Furthermore, just as before, it is assumed that there exists a positive number δ such that any two points $p, q \in M$ satisfying, for some $i = 0, 1, \ldots, s$, the inequalities

$$\rho(p, p_i) < \delta, \ \rho(q, p_{i+1}) < \delta,$$

where $p_j = \gamma(t_j)$, for $j = 0, 1, \ldots, s+1$, can be connected in M by a unique minimizing curve (cf. section 8). We assume that this number δ is so small that, for arbitrary $i = 1, \ldots, s$, the spherical δ-neighborhood of the point p_i is a coordinate neighborhood of that point, that is, so small that certain local coordinates

$$x_i^1, \ldots, x_i^m, \tag{1}$$

will be defined in it and the spherical δ-neighborhoods of "neighboring" points p_i and p_{i+1} will be disjoint for every $i = 1, \ldots, s$. We shall also assume that the local coordinates (1) vanish at the point p_i.

Let $V_i \subset R^m$ denote the image of the δ-neighborhood of the point p_i under the coordinate homeomorphism corresponding to the local coordinates (1) and define

$$V = V_1 \times \ldots \times V_s.$$

The set V is an open subset of the space R^{sm}, and each point $x = (x_i^k)$ in it, for $i = 1, \ldots, s$ and $k = 1, \ldots, m$, defines s points $q_i \in V_i$ at which

$$x_i^k(q_i) = x_i^k, \ i = 1, \ldots, s; \ k = 1, \ldots, m.$$

Since the point q_i belongs to the δ-neighborhood of the point p_i for every $i = 1, \ldots, s$, this point can be connected with the point q_{i+1} by a unique minimizing curve. (We assume that $q_{s+1} = p_{s+1} = \bar{q} = \gamma(1)$.) Furthermore, the point $\bar{p} = p_0 = \gamma(0)$ can be connected with the point

q_1 by a unique minimizing curve. Combining these two minimizing curves, we obtain a broken geodesic u^x connecting the points $\bar{p}$ and $\bar{q}$. We choose the parameter t on the geodesic u^x proportional to the arc length and specifically, in such a way that

$$u^x(0)=\bar{p}, \quad u^x(1)=\bar{q}.$$

We denote the value of the parameter t corresponding to the point q_i, for $i=0, \ldots s+1$, by $t_i(x)$, so that

$$u^x(t_i(x))=q_i.$$

In particular, $t_0(x)=0$ and $t_{s+1}(x)=1$.

We denote the length of the segment of the curve $u^x(t)$, for $t_i(x)\leqslant t\leqslant t_{i+1}(x)$, that is, the distance $\rho(q_i, q_{i+1})$ between the points q_i and q_{i+1}, by $J_i(x)$. This construction is completely analogous to the construction of the curves u^z described in section 8. Carrying this analogy further, let us consider, for an arbitrary point $x\in R^{sm}$, the variation φ_i^x defined by

$$\varphi_i^x(t, \tau)=u^{\tau x}(t), \quad |\tau|<\tau_0, \quad t_i(\tau x)\leqslant t\leqslant t_{i+1}(\tau x),$$

where τ_0 is a positive number such that $\tau x\in V$ for $|\tau|<\tau_0$. Then, proceeding word for word as in section 8, we obtain the following formula for the derivative of the function $J_i^x(\tau)=J_i(\tau x)$ expressing the length of the curve $u^{\tau x}(t)$ for $t_i(\tau x)\leqslant t\leqslant t_{i+1}(\tau x)$ [analogous to formula (6) of section 8]:

$$\dot{J}_i^x(0)=\frac{1}{l}\left[(\dot{a}_{i+1}^x(0), \dot{\gamma}_{i+1})-(\dot{a}_i^x(0), \dot{\gamma}_i)\right], \quad i=0, 1, \ldots, s, \tag{2}$$

where $l=k(0)$ is the length of the geodesic γ and $\dot{a}_j(0)$, for $j=0, 1, \ldots, s+1$, is the tangent vector to the broken curve

$$a_j^x(\tau)=u^{\tau x}(t_j(\tau x))$$

at the point $\tau=0$. Furthermore, still proceeding word for word as in section 10, we obtain the formula

$$\ddot{J}_i^x(0)=\frac{1}{l}\left[\left(\frac{\nabla\dot{a}_{i+1}^x}{dt}(0), \dot{\gamma}_{i+1}\right)-\left(\frac{\nabla\dot{a}_i^x}{dt}(0), \dot{\gamma}_i\right)\right]+ \\ +\frac{1}{l}I_{t_i}^{t_{i+1}}(X_i^x(t)), \tag{3}$$

where $X_i^x(t)$, for $t_i\leqslant t\leqslant t_{i+1}$, is the vector field associated with the variation φ_i^x.

At this point, we leave the analogy that we have been pursuing. Let us consider the function

$$J_*(x)=[J_1(x)]^2+\ldots+[J_s(x)]^2$$

and its Hessian

$$Q_*(x)=\frac{\partial^2 J_*(0)}{\partial x_{i_1}^{k_1}\partial x_{i_2}^{k_2}}x_{i_1}^{k_1}x_{i_2}^{k_2}$$

at the point $x=0$. We shall call this Hessian, which is a quadratic form defined on the space R^{sm}, *Bott's quadratic form.* In analogy with Morse's form,

$$Q_*(x)=\ddot{J}^x_*(0),$$

where

$$J^x_*(\tau)=J_*(\tau x)=[J^x_1(\tau)]^2+\ldots+[J^x_s(\tau)]^2.$$

Consequently,

$$Q_*(x)=2\sum_{i=0}^{s}([\dot{J}^x_i(0)]^2+J^x_i(0)\ddot{J}^x_i(0)). \qquad (4)$$

Let us now compare Bott's form $Q_*(x)$ and Morse's form $Q(z)$. To make this comparison, let us do the following:

First, let us suppose that the points $t_1, \ldots, t_s$ on which the construction of these forms depends are chosen *uniformly* on the interval [0, 1], that is, so that

$$t_1=\frac{1}{s+1}, \ldots, t_i=\frac{1}{s+1}, \ldots, t_s=\frac{s}{s+1},$$

where s is some sufficiently large number.

Second, let us suppose that the submanifolds $P_1, \ldots, P_s$ on which the construction of Morse's form $Q(z)$ depends are chosen so that, for arbitrary $i=1, \ldots, s$, the geodesic γ is orthogonal to the submanifold P_i, that is, chosen so that

$$(A_i, \dot{\gamma}_i)=0, \qquad i=1, \ldots, s,$$

for an arbitrary vector $A_i \in M_{p_i}$ tangent to the submanifold P_i.

Third, let us suppose that the coordinates (1) on which the construction of Bott's form $Q_*(x)$ depends are chosen so that, for arbitrary $i=1, \ldots, s$, the submanifold P_i is determined by the equation $x^m_i=0$ and that the restrictions

$$z^1_i=x^1_i|_{P_i}, \ldots, z^{m-1}_i=x^{m-1}_i|_{P_i} \qquad (5)$$

to the P_i of the coordinates $x^1_i, \ldots, x^{m-1}_i$ constitute a system of local coordinates on the P_i. Let us suppose also that, in a δ-neighborhood of the point p_i, the geodesic γ is defined by the equations

$$x^1_i=0, \ldots, x^{m-1}_i=0$$

(cf. end of section 14, Chapter 1).

Fourth, let us take the coordinates (5) for the coordinates z^k_i, where $i=1, \ldots, s$ and $k=1, \ldots, m-1$, that were mentioned in the construction of Morse's form $Q(z)$.

Fifth, let us take for the space $R^{s(m-1)}$ on which Morse's form $Q(z)$ is defined the subspace

$$x^m_1=0, \ldots, x^m_s=0$$

of the space R^{sm} on which Bott's form $Q_*(x)$ is defined.

Under these assumptions, Morse's form Q(z) differs only by a constant factor from the restriction $Q_*(z)$ *of Bott's form* $Q_*(x)$ *to the subspace* $R^{s(m-1)}$.

Proof: Since for arbitrary $i=1, \ldots, s$, the vector $\dot{a}_i^z(0)$, where $z \in R^{s(m-1)}$ is a tangent vector of the submanifold P_i and, hence, by hypothesis, orthogonal to the vector $\dot{\gamma}_i$ and since the vectors $\dot{a}_0^z(0)$ and $\dot{a}^z_{s+1}(0)$ are equal to 0, we have, in accordance with formula (6) of section 8,

$$\dot{j}_i^z(0)=0, \quad i=0, 1, \ldots, s.$$

Furthermore, owing to the uniform distribution of the points t_i on the interval [0, 1], we have

$$J_i^z(0)=\frac{1}{s+1}l$$

for arbitrary $i=0, 1, \ldots, s$. Finally, the functions $J_i^x(\tau)$ introduced above obviously coincide at $x=z$ with the functions $J_i^z(\tau)$, considered in section 10. Since

$$Q(z)=\sum_{i=0}^{s}\ddot{j}_i^z(0),$$

it follows from this and from formula (4) that

$$Q_*(z)=\frac{2l}{s+1}Q(z)$$

for an arbitrary point $x=z \in R^{s(m-1)}$.

Remark: Since

$$\dot{j}_*^x(0)=2\sum_{i=0}^{s}J_i^x(0)\dot{j}_i^x(0)=\frac{2l}{s+1}\sum_{i=0}^{s}\dot{j}_i^x(0), \quad x \in R^{sm},$$

it follows from formula (2) [by virtue of the equations $\dot{a}_0^x(0)=0$ and $\dot{a}_{s+1}^x(0)=0$] that

$$\dot{j}_*^x(0)=0$$

for an arbitrary point $x \in R^{sm}$. Since

$$\dot{j}_*^x(0)=\left(\frac{\partial J_*(x)}{\partial x_i^k}\right)_{x=0}x_i^k,$$

it follows from this that

$$\left(\frac{\partial J_*(x)}{\partial x_i^k}\right)_{x=0}=0. \tag{6}$$

Let us now look at the form $Q_*(x)$ on the subspace R^s of the space R^{ms} complementary to the subspace $R^{s(m-1)}$, that is, consisting of those points $x \in R^{sm}$ of which only the coordinates $x_1^m, \ldots, x_s^m$ are nonzero. For every point $x \in R^s$ and every $\tau \in [-\tau_0, \tau_0]$, the curve $u^{\tau x}(t)$ obviously coincides with the geodesic $\gamma(t)$, so that

$$a_i^x(\tau) = \gamma(t_i^x(\tau)),$$

where $t_i^x(\tau) = t_i(\tau x)$ and hence

$$\dot{a}_i^x(0) = \dot{\gamma}_i \dot{t}_i^x(0),$$
$$\frac{\nabla \dot{a}_i^\tau}{d\tau}(0) = \dot{\gamma}_i \ddot{t}_i^x(0)$$

$\left(\text{since } \frac{\nabla \dot{\gamma}}{dt}(t) = 0\right)$. Furthermore, $X_i^x(t) \equiv 0$. Consequently,

$$\dot{J}_i^x(0) = \dot{t}_{i+1}^x(0) - \dot{t}_i^x(0),$$
$$\ddot{J}_i^x(0) = \ddot{t}_{i+1}^x(0) - \ddot{t}_i^x(0)$$

(cf. formulas (2) and (3) and recall that $(\dot{\gamma}_i, \dot{\gamma}_i) = l$ for arbitrary i). Therefore, in accordance with formula (4),

$$Q_*(x) = 2\sum_{i=0}^{s}(\dot{t}_{i+1}^x(0) - \dot{t}_i^x(0))^2 + $$
$$+ \frac{2l}{s+1}\sum_{i=0}^{s}(\ddot{t}_{i+1}^x(0) - \ddot{t}_i^x(0)) = 2\sum_{i=0}^{s}(\dot{t}_{i+1}^x(0) - \dot{t}_i^x(0))^2,$$

since $\ddot{t}_0^x(0) = \ddot{t}_{s+1}^x(0) = 0$ (because $t_0^x(\tau) = 0$ and $t_{s+1}^x(\tau) = 1$ for arbitrary τ). Consequently,

$$Q_*(x) \geqslant 0,$$

with equality holding only when $\dot{t}_{i+1}^x(0) = \dot{t}_i^x(0)$ for arbitrary $i = 0, 1, \ldots, s$, that is, only when $\dot{t}_i^x(0) = 0$ for all $i = 1, \ldots, s$ (since, for example, $\dot{t}_0^x(0) = 0$ for all x). But, by definition, the number $t_i(x)$ corresponds on the geodesic $\gamma(t)$ to the point $q_i = \gamma(t_i(x))$, for which $x_i^k(q_i) = x_i^k$. In particular,

$$x_i^m(\gamma(t_i(x))) = x_i^m$$

that is,

$$x_i^m(t_i(x)) = x_i^m,$$

where $x_i^m(t) = x_i^m(\gamma(t))$. If we differentiate this equation with respect to x_i^m and set $x = 0$, we obtain the relation

$$\dot{x}_i^m(t_i)\left(\frac{\partial t_i(x)}{\partial x_i^m}\right)_{x=0} = 1.$$

(Here, and in the following formulas, we are *not* using the summation convention with respect to i.) Consequently,

$$\left(\frac{\partial t_i(x)}{\partial x_i^m}\right)_{x=0} \neq 0.$$

On the other hand, if we differentiate the function $t_i^x(\tau) = t_i(\tau x)$ with respect to τ and set $\tau = 0$, we obtain

$$\dot{t}_i^x(0) = \sum_{j=1}^{s} \left(\frac{\partial t_i(x)}{\partial x_j^m}\right)_{x=0} \cdot x_j^m = \left(\frac{\partial t_i(x)}{\partial x_i^m}\right)_{x=0} x_i^m,$$

since the function $t_i(x)$ obviously depends only on the variable x_i^m (we recall that $x \in R^m$ and hence $x_j^k = 0$ for $k \neq m$). Thus,

If $x_i^m \neq 0$, then $\dot{t}_i^x(0) \neq 0$ for $i = 1, \ldots, s$.

In other words, the equation $\dot{t}_i^x(0) = 0$ for all $i = 1, \ldots, s$ is possible only when $x = 0$. Thus, we have shown that

The quadratic form $Q_*(x)$ is positive-definite on the subspace R^s of the space R^{sm}.

Let us show now that

The pair $(R^s, R^{s(m-1)})$ implements a decomposition of the form $Q_*(x)$ as a direct sum; that is,

$$Q_*(x + z) = Q_*(x) + Q_*(z)$$

for arbitrary points $x \in R^s$ and $z \in R^{s(m-1)}$.

Proof: (Cf. the reasoning in section 9.) One can easily see that the mapping

$$R^{sm} \to M_{p_i},$$

which assigns to each point $x \in R^{sm}$ the vector $\dot{a}_i^x(0) \in M_{p_i}$ is linear. In particular,

$$\dot{a}_i^{x+z}(0) = \dot{a}_i^x(0) + \dot{a}_i^z(0)$$

for arbitrary points $x \in R^s$ and $z \in R^{s(m-1)}$. Since, by hypothesis, $(\dot{a}_i^z(0), \dot{\gamma}_i) = 0$, it follows on the basis of formula (2) that

$$\dot{t}_i^{x+z}(0) = \dot{t}_i^x(0)$$

for arbitrary points $x \in R^s$ and $z \in R^{s(m-1)}$.

Analogously, the mapping

$$R^{sm} \to R,$$

which assigns to every point $x \in R^{sm}$ the number $\dot{t}_i^x(0)$ is also linear. Therefore, by virtue of the formulas

$$X_i^x(t_i) = \dot{a}_i^x(0) - \dot{t}_i^x(0)\,\dot{\gamma}_i,$$

$$X_i^x(t_{i+1}) = \dot{a}_{i+1}^x(0) - \dot{t}_{i+1}^x(0)\,\dot{\gamma}_{i+1}$$

and the fact that the field $X_i^x(t)$ on the interval $[t_i, t_{i+1}]$ is unambiguously defined by the vectors $X_i^x(t_i)$ and $X_i^x(t_{i+1})$ and depends linearly on them, the mapping that assigns to each point $x \in R^{sm}$ the field $X_i^x(t)$ is also linear. In particular,

$$X_i^{x+z}(t) = X_i^z(t)$$

for arbitrary points $x \in R^s$ and $z \in R^{s(m-1)}$ (since, as we know, $X_i^x(t) \equiv 0$). Consequently [cf. formula (3)],

$$\ddot{j}_i^{x+z}(0) = \frac{1}{l}\left[\left(\frac{\nabla \dot{a}_{i+1}^{x+z}}{d\tau}(0), \dot{\gamma}_{i+1}\right) - \left(\frac{\nabla \dot{a}_i^{x+z}}{d\tau}(0), \dot{\gamma}_i\right)\right] + \\ + \frac{1}{l} I_{t_i}^{t_{i+1}}(X^z(t)),$$

and, therefore,

$$\sum_{i=0}^{s} \ddot{j}_i^{x+z}(0) = \frac{1}{l} I_0^1(X^z(t)) = Q(z),$$

since $\frac{\nabla \dot{a}_0^{x+z}}{d\tau}(0) = 0$ and $\frac{\nabla \dot{a}_{s+1}^{x+z}}{d\tau}(0) = 0$.

Thus,

$$\begin{aligned} Q_*(x+z) &= 2\sum_{i=0}^{s}\left([\dot{j}_i^{x+z}(0)]^2 + j_i^{x+z}(0)\ddot{j}_i^{x+z}(0)\right) = \\ &= 2\sum_{i=0}^{s}(\dot{j}_i^x(0))^2 + \frac{2l}{s+1}\sum_{i=0}^{s}\ddot{j}_i^{x+z}(0) = \\ &= Q_*(x) + \frac{2l}{s+1}Q(z) = \\ &= Q_*(x) + Q_*(z), \end{aligned}$$

as asserted.

It follows immediately from the assertions proven that

The corank and negative index of inertia of Bott's form $Q_*(x)$ *are equal, respectively, to the corank and negative index of inertia of Morse's form* $Q(z)$.

Thus, we can now formulate the theorems on indices that we have proved in sections 11 and 12 in the following form:

The index $\lambda_1(0)$ *of the point* $t = 0$ *(with respect to the point* $t = 1$*) is equal to the corank of Bott's form* $Q_*(z)$, *and the index* $\lambda_1(0, 1)$ *of the interval (0, 1) is equal to its negative index of inertia.*

APPENDIX

FOCAL POINTS

1. The second quadratic form of a submanifold

All the results that we have expounded above refer to two fixed points $\bar{p}$ and $\bar{q}$ in the space M and to some geodesic $\gamma(t)$ passing through these points. Now, we shall generalize these results to the case in which the point $\bar{q}$ is replaced with some submanifold $N \subset M$ and the geodesic $\gamma(t)$ [which, as before, passes through the point $\bar{p}$] is orthogonal to the submanifold N. To do this, we need, as a preliminary, to generalize the concept (familiar from elementary differential geometry) of the second quadratic form of a surface in three-dimensional Euclidean space to the case of arbitrary submanifolds of arbitrary Riemannian spaces.

Let M denote a Riemannian space, let N denote a submanifold of it, let p denote a point in the submanifold N, and let $N_p \subset M_p$ denote the tangent space of the submanifold N at the point p. Let us choose two vectors A and B in the space N_p and let us consider an arbitrary curve $\gamma(t)$ on the submanifold N such that

$$\gamma(0) = p, \qquad \dot{\gamma}(0) = A.$$

Let us choose on this curve a field $Y_*(t)$ of vectors $Y_*(t) \in N_{\gamma(t)}$ such that

$$Y_*(0) = B.$$

(To construct such a field, we may, for example, make a parallel translation of the vector B into each point of the curve $\gamma(t)$, treated as a curve in M, and then project the resulting vectors $Y(t) \in M_{\gamma(t)}$ orthogonally onto the subspaces $N_{\gamma(t)}$.) If we differentiate the field $Y_*(t)$ covariantly along the curve $\gamma(t)$, we obtain a field

$$\frac{\Delta Y_*}{dt}(t) \in M_{\gamma(t)}$$

on that curve. In particular, for $t = 0$, we obtain the vector

$$\frac{\Delta Y_*}{dt}(0) \in M_p.$$

Now, let C denote an arbitrary vector (in the space M_p) that is orthogonal to the submanifold N (i.e., orthogonal to the subspace N_p). We set

$$\pi_C(A, B) = \left(C, \frac{\Delta Y_*}{dt}(0)\right).$$

Then,

The number $\pi_C(A, B)$ *is independent of the choice of the curve* $\gamma(t)$ *and the field* $Y_*(t)$; *that is, it is determined exclusively by the vectors* $A, B \in N_p$ *and the vector* $C \perp N_p$.

Proof: Let

$$x^1, \ldots, x^n$$

denote an arbitrary system of local coordinates of the manifold N at the point p that are equal to 0 at that point and let

$$\xi : U \to R^n, \qquad U \subset N,$$

be the corresponding coordinate homeomorphism. The inverse homeomorphism $\varphi = \xi^{-1}$: $U^0 \to N$, where $U^0 = \xi U \subset R^n$, considered as the mapping $U^0 \to M$ can be treated as an n-dimensional surface in M defined on an open set U^0. The basis vectors

$$\frac{\partial \varphi}{\partial t^i}(t^1, \ldots, t^n), \qquad (t^1, \ldots, t^n) \in U^0,$$

of this space are connected with the vectors $\partial / \partial x^i$ by the formulas

$$\frac{\partial \varphi}{\partial t^i}(x^1(q), \ldots, x^n(q)) = \left(\frac{\partial}{\partial x^i}\right)_q, \qquad q \in U.$$

In particular,

$$\frac{\partial \varphi}{\partial t^i}(0, \ldots, 0) = \left(\frac{\partial}{\partial x^i}\right)_p, \qquad i = 1, \ldots, n.$$

The curve $\gamma(t)$ belongs to the surface φ and is defined on it by the functions $x^i(t) = x^i(\gamma(t))$; that is,

$$\varphi(x^1(t), \ldots, x^n(t)) = \gamma(t).$$

Consequently, if we set

$$Y_*(t) = Y_*^j(t)\left(\frac{\partial}{\partial x^j}\right)_{\gamma(t)} = Y_*^j(t)\frac{\partial \varphi}{\partial t^j}(x^1(t), \ldots, x^n(t)),$$

(summing with respect to j from 1 to n), we obtain

$$\frac{\nabla Y_*}{dt}(t) = \frac{dY_*^j(t)}{dt}\frac{\partial \varphi}{\partial t^j}(x^1(t), \ldots, x^n(t)) +$$

$$+ Y_*^j(t)\frac{\nabla}{dt}\left(\frac{\partial \varphi}{\partial t^j}(x^1(t), \ldots, x^n(t))\right) =$$

$$= \frac{dY_*^j(t)}{dt}\left(\frac{\partial}{\partial x^j}\right)_{\gamma(t)} + Y_*^j(t)\dot{x}^i(t)\left(\frac{\nabla}{\partial t^i}\frac{\partial \varphi}{\partial t^j}\right)(x^1(t), \ldots, x^n(t))$$

(cf. formula (4) of section 3, Chapter 2). In particular,

$$\frac{\nabla Y_*}{dt}(0) = \frac{dY_*^j}{dt}(0)\left(\frac{\partial}{\partial x^j}\right)_p + A^i B^j \left(\frac{\nabla}{\partial t^i}\frac{\partial \varphi}{\partial t^j}\right)(0, \ldots, 0),$$

where $A^i = \dot{x}^i(0)$ and $B^j = Y_*^j(0)$ are the components of the vectors A and B, respectively, relative to the basis

$$\left(\frac{\partial}{\partial x^1}\right)_p, \ldots, \left(\frac{\partial}{\partial x^n}\right)_p.$$

Since $C \perp N_p$ and $\left(C, \left(\frac{\partial}{\partial x^i}\right)_p\right) = 0$, for all $i = 1, \ldots, n$, it follows that

$$\pi_C(A, B) = \left(C, \frac{\nabla}{\partial t^i}\frac{\partial \varphi}{\partial t^j}(0, \ldots, 0)\right) A^i B^j.$$

This completes the proof of the assertion made above since the right-hand member of this formula is independent of the choice of the curve $\gamma(t)$ and the field $Y_*(t)$.

Furthermore, it follows from the formula that we have proved and formula (2) of section 3 of Chapter 2, that

The function $\pi_C(A, B)$ *of the vectors* A, $B \in N_p$ *is a symmetric bilinear form on the space* N_p.

The corresponding quadratic form

$$\pi_C(A) = \pi_C(A, A)$$

is a precise analog of the second quadratic form of surfaces in three-dimensional Euclidean space. It follows immediately from what was said above that

$$\pi_C(A) = \left(\frac{\nabla \dot{\gamma}}{dt}(0), C\right),$$

where $\gamma(t)$ is an arbitrary curve on the submanifold N such that

$$\gamma(0) = p, \quad \dot{\gamma}(0) = A.$$

In what follows, it will be more convenient for us to consider not the form $\pi_C(A)$ but the corresponding self-adjoint operator T_C: $N_p \to N_p$. To obtain a direct geodetric definition of this operator, let us consider on the curve $\gamma(t)$ the field of parallel vectors $Z(t)$ for which

$$Z(0) = C.$$

Let $Z_*(t) \in N_{\gamma(t)}$ be the orthogonal projection of the vector $Z(t) \in M_{\gamma(t)}$ onto the subspace $N_{\gamma(t)}$. We obtain

$$T_C(A) = \frac{\nabla Z_*}{dt}(0).$$

Let us show that

The vector $T_C(A)$ belongs to the subspace N_p and is independent of the choice of the curve $\gamma(t)$.

To show this, let us set

$$Z_*(t) = Z_*^j(t)\left(\frac{\partial}{\partial x^j}\right)_{\gamma(t)}.$$

Just as before, we obtain

$$\frac{\nabla Z_*}{dt}(t) = \frac{dZ_*^j(t)}{dt}\left(\frac{\partial}{\partial x^j}\right)_{\gamma(t)} + $$
$$+ Z_*^j(t)\,\dot{x}^i(t)\left(\frac{\nabla}{\partial t^i}\frac{\partial \varphi}{\partial t^j}\right)(x^1(t), \ldots, x^n(t)).$$

On the other hand, since $C \perp N_p$, it follows that $Z_*^j(0) = 0$ for all $j = 1, \ldots, n$. Consequently,

$$\frac{\nabla Z_*}{dt}(0) = \frac{dZ_*^j(0)}{dt}\left(\frac{\partial}{\partial x^j}\right)_{\gamma(t)}$$

and, hence,

$$T_C A = \frac{\nabla Z_*}{dt}(0) \in N_p.$$

Furthermore, let $Y(t)$ denote an arbitrary field of parallel vectors on the curve $\gamma(t)$ and let $Y_*(t)$ denote the orthogonal projection of the vector $Y(t)$ onto the subspace $N_{\gamma(t)}$. Define

$$Y_0(t) = Y(t) - Y_*(t), \qquad Z_0(t) = Z(t) - Z_*(t).$$

Since $Y_0(t) \perp N_{\gamma(t)}$ and $Z_0(t) \perp N_{\gamma(t)}$,

$$(Z_*(t), Y_0(t)) = 0, \qquad (Z_0(t), Y_*(t)) = 0.$$

If we differentiate these relations with respect to t, we obtain

$$\left(\frac{\nabla Z_*}{dt}(t), Y_0(t)\right) + \left(Z_*(t), \frac{\nabla Y_0}{dt}(t)\right) = 0,$$
$$\left(\frac{\nabla Z_0}{dt}(t), Y_*(t)\right) + \left(Z_0(t), \frac{\nabla Y_*}{dt}(t)\right) = 0.$$

But

$$\frac{\nabla Y_0}{dt}(t) = -\frac{\nabla Y_*}{dt}(t),$$
$$\frac{\nabla Z_0}{dt}(t) = -\frac{\nabla Z_*}{dt}(t),$$

since $\frac{\nabla Y}{dt}(t) = 0$ and $\frac{\nabla Z}{dt}(t) = 0$. Consequently,

$$\left(\frac{\nabla Z_*}{dt}(t), Y_0(t)\right) = \left(Z_*(t), \frac{\nabla Y_*}{dt}(t)\right),$$
$$\left(\frac{\nabla Z_*}{dt}(t), Y_*(t)\right) = \left(Z_0(t), \frac{\nabla Y_*}{dt}(t)\right).$$

Adding these equations, we obtain the equation

$$\left(\frac{\nabla Z_*}{dt}(t), Y(t)\right) = \left(Z(t), \frac{\nabla Y_*}{dt}(t)\right).$$

If we set $t = 0$ in this equation, we obtain the relation

$$(T_C A, B) = \pi_C(A, B), \tag{1}$$

where $B = Y(0)$. Since $\pi_C(A, B)$ is independent of the choice of the curve $\gamma(t)$, this equation shows that, for an arbitrary vector $B \in N_p$, the scalar product $(T_C A, B)$ is also independent of the choice of that curve. Since $T_C A \in N_p$, this is possible only when the vector $T_C A$ is independent of the choice of curve $\gamma(t)$. This completes the proof of the assertion made above.

Furthermore, it follows from formula (1) that

The mapping $T_C: N_p \to N_p$ *is linear and self-adjoint; that is,*

$$T_C(\alpha A + \beta B) = \alpha T_C A + \beta T_C B$$

and

$$(T_C A, B) = (A, T_C B)$$

for arbitrary vectors A, $B \in N_p$ *and arbitrary numbers* α *and* β.

2. Focal points

Now, let $\gamma(t)$ denote a geodesic such that, for some $t_1 \in R$,

$$\gamma(t_1) \in N, \quad \dot{\gamma}(t_1) \perp N_{\gamma(t_1)}.$$

Let us consider the subspace $\Lambda_\gamma^N(t_1)$ of the space J_γ of Jacobi fields that consists of those Jacobi fields $X(t)$ on the geodesic $\gamma(t)$ such that

$$X(t_1) \in N_{\gamma(t_1)},$$
$$\frac{\nabla X}{dt}(t_1) + T_{\dot{\gamma}(t_1)} X(t_1) \perp N_{\gamma(t_1)}. \tag{1}$$

Relations (1) are linear and they impose on the solutions $X(t)$ of equation (3) of section 3 of Chapter 4 exactly m independent conditions. Consequently,

The dimension of the subspace $\Lambda_\gamma^N(t_1)$ *is equal to the dimension* m *of the manifold M.*

We call a point $p = \gamma(t_0)$ of the geodesic $\gamma(t)$ (or the corresponding point $t_0 \in R$) a *focal point* of the submanifold N if there exists a nonzero Jacobi field $X(t) \in \Lambda_\gamma^N(t_1)$ such that

$$X(t_0) = 0,$$

that is, if the subspace

$$\Lambda_\gamma(t_0) \cap \Lambda_\gamma^N(t_1)$$

of the space J_γ is nonzero. We call the dimension $\lambda_N(t_0)$ of this subspace the index of the point p (or the point t_0) relative to the submanifold N. Thus,

A point t_0 is a focal point of a submanifold N if and only if its index $\lambda_N(t_0)$ is nonzero.

The definition of focal points is completely analogous to the definition of conjugate points and is obtained from the latter definition by replacing the subspace $\Lambda_\gamma(t_1)$ with the subspace $\Lambda_\gamma^N(t_1)$. Let us pursue this analogy in greater detail.

In the first place, it is easy to see that

For any two fields $X(t), Y(t) \in \Lambda_\gamma^N(t_1)$,

$$\left(\frac{\nabla X}{dt}(t),\ Y(t)\right) = \left(X(t),\ \frac{\nabla Y}{\partial t}(t)\right). \tag{2}$$

(This equation is analogous to equation (5), section 3, Chapter 4.)

Proof: As we know (cf. proof of equation (5), section 3, Chapter 4), for any two fields $X(t), Y(t) \in J_\gamma$, we have

$$\left(\frac{\nabla X}{dt}(t),\ Y(t)\right) - \left(X(t),\ \frac{\nabla Y}{dt}(t)\right) = \text{const.} \tag{2'}$$

But, if $X(t), Y(t) \in \Lambda_\gamma^N(t_1)$, then

$$\left(\frac{\nabla X}{dt}(t_1),\ Y(t_1)\right) - \left(X(t_1),\ \frac{\nabla Y}{dt}(t_1)\right) =$$
$$= -\left(T_{\dot\gamma(t_1)} X(t_1),\ Y(t_1)\right) + \left(X(t_1),\ T_{\dot\gamma(t_1)} Y(t_1)\right) = 0,$$

since the operator $T_{\dot\gamma(t_1)}$ is self-adjoint. Consequently, the constant in formula (2′) is equal to 0.

Analogously, formula (6) of section 3 of Chapter 4 remains valid in the present case; that is,

For an arbitrary field $X(t) \in \Lambda_\gamma(t_0) \cap \Lambda_\gamma^N(t_1)$

$$(X(t),\ \dot\gamma(t)) = 0. \tag{3}$$

Proof of formula (3) coincides word for word with the proof of formula (6) of section 3 of Chapter 4. (One need only remember that, by hypothesis, $\dot\gamma(t_1) \perp X(t_1)$.)

The construction of fields in $\Lambda_\gamma^N(t_1)$ with the aid of Jacobian variations is described by the following proposition:

A Jacobi field $X(t)$ on a geodesic $\gamma(t)$ belongs to the subspace $\Lambda_\gamma^N(t_1)$ if and only if it is associated with a Jacobian variation $\varphi(t, \tau)$ of the geodesic $\gamma(t)$ such that

$$\begin{aligned} &\varphi(t_1, \tau) \in N \quad \textit{for all} \quad \tau, \\ &\frac{\partial \varphi}{\partial t}(t_1, \tau) \perp N_{\varphi(t, \tau)} \quad \textit{for all} \quad \tau. \end{aligned} \tag{4}$$

Suppose that $\varphi(t, \tau)$ satisfies conditions (4). These conditions mean, in particular, that the curve

$$\alpha(\tau) = \varphi(t_1, \tau)$$

belongs to the submanifold N. Therefore,

$$\dot{\alpha}(\tau) \in N_{\alpha(\tau)}$$

and, in particular,

$$X(t_1) = \dot{\alpha}(0) \in N_{\gamma(t_1)}.$$

Furthermore, conditions (4) also indicate that the vector field

$$\mathrm{A}(\tau) = \frac{\partial \varphi}{\partial t}(t_1, \tau)$$

on the curve $\alpha(\tau)$ is orthogonal to N.

Let A denote an arbitrary vector in the space $N_{\gamma(t_1)}$. By making a parallel translation of the vector A along the curve $\alpha(\tau)$ and projecting the resulting vectors orthogonally onto the subspace $N_{\alpha(\tau)}$, we obtain on the curve $\alpha(\tau)$ a vector field $A_*(\tau)$. Since $\mathrm{A}(\tau) \perp N_{\alpha(\tau)}$, we have

$$(\mathrm{A}(\tau), A_*(\tau)) = 0 \quad \text{for all} \quad \tau.$$

If we differentiate this equation with respect to τ and then set $\tau = 0$, we obtain, by virtue of the relation

$$\frac{\nabla \mathrm{A}}{d\tau}(0) = \frac{\nabla}{\partial \tau} \frac{\partial \varphi}{\partial t}(t_1, 0) = \frac{\nabla}{\partial t} \frac{\partial \varphi}{\partial \tau}(t_1, 0) = \frac{\nabla X}{\partial t}(t_1),$$

the equation

$$\left(\frac{\nabla X}{dt}(t_1), A\right) + \left(\dot{\gamma}(t_1), \frac{\nabla A_*}{d\tau}(0)\right) = 0.$$

Since $\dot{\gamma}(t_1) \perp N_{\gamma(t_1)}$ and $\dot{\alpha}(0) = X(t_1)$, the second term in this expression is equal to

$$\pi_{\dot{\gamma}(t_1)}(X(t_1), A) = \left(T_{\dot{\gamma}(t_1)} X(t_1), A\right).$$

Therefore,

$$\left(\frac{\nabla X}{dt}(t_1) + T_{\dot{\gamma}(t_1)} X(t_1), A\right) = 0,$$

that is,

$$\frac{\nabla X}{dt}(t_1) + T_{\dot{\gamma}(t_1)} X(t_1) \perp N_{\gamma(t_1)}.$$

Thus, $X(t) \in \Lambda_\gamma^N(t_1)$.

Conversely, let $X(t)$ denote an arbitrary Jacobi field belonging to the subspace $\Lambda_\gamma^N(t_1)$. Since $X(t_1) \in N_{\gamma(t_1)}$, there exists on the manifold N a curve $\alpha(\tau)$ such that

$$\alpha(0) = \gamma(t_1), \quad \dot{\alpha}(0) = X(t_1).$$

Let us construct on the curve $\alpha(\tau)$ an arbitrary field $Y(\tau)$ such that

$$Y(0) = \dot{\gamma}(t_1), \quad \frac{\nabla Y}{d\tau}(0) = \frac{\nabla X}{dt}(0),$$

and then let us set

$$Y(\tau) = \mathrm{A}(\tau) + \mathrm{B}(\tau),$$

where $\mathrm{A}(\tau) \perp N_{\alpha(\tau)}$ and $\mathrm{B}(\tau) \in N_{\alpha(\tau)}$. Since $Y(0) \perp N_{\gamma(t_1)}$, we have

$$\mathrm{A}(0) = \dot{\gamma}(t_1).$$

Consequently, conditions (1) of section 3 of Chapter 4 (with t_0 replaced by t_1) are satisfied for the curve $\alpha(\tau)$ and the field $\mathrm{A}(\tau)$. Therefore, there exists a Jacobian variation $\varphi(t, \varphi)$ of the geodesic $\gamma(t)$ such that

$$\varphi(t_1, \tau) = \alpha(\tau), \quad \frac{\partial \varphi}{\partial t}(t_1, \tau) = \mathrm{A}(\tau) \quad \text{for all } \tau.$$

Since this variation satisfies conditions (4), to complete the proof, it remains only to show that the Jacobi field associated with it

$$\bar{X}(t) = \frac{\partial \varphi}{\partial \tau}(t, 0)$$

coincides with the field $X(t)$. To prove this in turn, we need only show that

$$\begin{aligned} \bar{X}(t_1) &= X(t_1), \\ \frac{\nabla \bar{X}}{dt}(t_1) &= \frac{\nabla X}{dt}(t_1). \end{aligned} \tag{5}$$

But, by construction,

$$\bar{X}(t_1) = \frac{\partial \varphi}{\partial \tau}(t_1, 0) = \dot{\alpha}(0) = X(t_1).$$

Analogously,

$$\begin{aligned} \frac{\nabla \bar{X}}{dt}(t_1) &= \frac{\nabla}{\partial t}\frac{\partial \varphi}{\partial \tau}(t_1, 0) = \frac{\nabla}{\partial \tau}\frac{\partial \varphi}{\partial t}(t_1, 0) = \\ &= \frac{\nabla \mathrm{A}}{d\tau}(0) = \frac{\nabla Y}{d\tau}(0) - \frac{\nabla \mathrm{B}}{d\tau}(0) = \frac{\nabla X}{dt}(t_1) - \frac{\nabla \mathrm{B}}{d\tau}(0). \end{aligned}$$

Thus, we need only show that

$$\frac{\nabla B}{d\tau}(0) = 0.$$

Since the variation $\varphi(t, \tau)$ satisfies conditions (4), we have, by virtue of what we have proved,

$$\frac{\nabla \bar{X}}{dt}(t_1) + T_{\dot{\gamma}(t_1)} \bar{X}(t_1) \perp N_{\gamma(t_1)},$$

and, hence,

$$\frac{\nabla \bar{X}}{dt}(t_1) - \frac{\nabla X}{dt}(t_1) \perp N_{\gamma(t_1)}, \tag{6}$$

because $\bar{X}(t_1) = X(t_1)$ and, by hypothesis,

$$\frac{\nabla X}{dt}(t_1) + T_{\dot{\gamma}(t_1)} X(t_1) \perp N_{\gamma(t_1)}.$$

Equation (6) means that

$$\frac{\nabla B}{d\tau}(0) \perp N_{\gamma(t_1)}.$$

Consequently, it suffices to show that

$$\left(\frac{\nabla B}{d\tau}(0), C\right) = 0$$

for an arbitrary vector $C \perp N_{\gamma(t_1)}$.

Let $Z(\tau)$ denote a field of parallel vectors in the curve $\alpha(\tau)$ such that

$$Z(0) = C,$$

and set

$$Z(\tau) = Z_*(\tau) + Z_0(\tau),$$

where $Z_*(\tau) \in N_{\alpha(\tau)}$ and $Z_0(\tau) \perp N_{\alpha(\tau)}$. If we differentiate the equation

$$(B(\tau), Z_0(\tau)) = 0$$

with respect to τ and set $\tau = 0$, we obtain

$$\left(\frac{\nabla B}{d\tau}(0), C\right) = 0,$$

since $Z_0(0) = C$ and $B(0) = 0$.

This completes the proof of the proposition stated above.

In section 4 of Chapter 4, we derived from the corresponding proposition regarding the space $\Lambda_\gamma(t_1)$ the result that conjugate points cannot be too close to each other. Analogously, we can now show that a point p sufficiently close to the submanifold N is not a focal point for any geodesic γ passing through that point. However, we shall not need this fact and we leave it unproven.

Another important property of conjugate points, namely, their discreteness, carries over to focal points. In other words,

The focal points of a submanifold N (corresponding to a given geodesic $\gamma(t)$) are discrete on the real axis, so that in any (open or closed) interval I there are only finitely many of these points.

The proof of this proposition repeats word for word (with consideration of the remark made at the end of section 4 of Chapter 4) the proof of the corresponding proposition for conjugate points (we need only replace the reference to formula (5) of section 3 of Chapter 4 with a reference to formula (2) of the present section).

In analogy with the case of conjugate points, we shall call the sum $\lambda_N(I)$ of the indices of focal points contained in an interval I the index of that interval (with respect to the submanifold N).

3. Evaluation of the index λ_N

The analogy between conjugate and focal points enables us to carry the results of section 4 dealing with the evaluation of the indices of points and intervals over to the case of focal points. Now, we shall show how, actually, this can be done.

To do this, let us first carry over the results of sections 6 and 7 of Chapter 4 dealing with the functional I_a^b to the case of the more general functional

$$\overline{I}_a^b(X(t)) = (TX(b),\ X(b)) + I_a^b(X(t)),\quad X(t) \in \mathfrak{M},$$

where T is an arbitrary self-adjoint operator $M_{\gamma(b)} \rightarrow M_{\gamma(b)}$.

Let $X(t)$ and $Y(t)$ denote two fields in $\mathfrak{M} = \mathfrak{M}_\gamma$ that assume the same values at the points a and b:

$$X(a) = Y(a),\qquad X(b) = Y(b). \tag{1}$$

Obviously, the inequality

$$\overline{I}_a^b(X(t)) \leqslant \overline{I}_a^b(Y(t)) \tag{2}$$

holds if and only if

$$I_a^b(X(t)) \leqslant I_a^b(Y(t)).$$

Consequently,

The minimal fields in the sense of section 6 of Chapter 4 are also minimal fields with respect to the functional $\overline{I}_a^b$; that is, inequality (2) holds for an arbitrary such field $X(t)$ and an arbitrary field $Y(t) \in \mathfrak{M}_\gamma$ that satisfies conditions (1).

In particular (cf. section 6 of Chapter 4),

An arbitrary minimal field with respect to the functional $\overline{I}_a^b$ is a Jacobi field.

Furthermore (cf. section 7, Chapter 4),

If an interval [a, b] contains no points conjugate to the point b, then

$$\bar{I}_a^b(X(t)) \leqslant \bar{I}_a^b(Y(t)), \tag{3}$$

where X(t) is a Jacobi field such that

$$X(a) = Y(a), \quad X(b) = Y(b)$$

for an arbitrary nonzero field $Y(t) \in \mathfrak{M}$. *Equality holds in formula (3) only when Y(t) = X(t).*

Of course, the analog of inequality (2) of section 7 of Chapter 4 for the functional $\bar{I}_a^b$ also holds. However, as will be clear from what follows, this inequality is not sufficient in the present case. We need the following stronger assertion:

If a point a is sufficiently close to a point b, then

$$\bar{I}_a^b(X(t)) > 0 \tag{4}$$

for an arbitrary nonzero field $X(t) \in \mathfrak{M}_\gamma$ *that vanishes at the point a.*

This assertion is stronger than the previous one since we do not now require that $X(b) = 0$.

In order not to interrupt the exposition, we postpone proof of this inequality until section 4.

Let us now turn to the construction of broken Jacobi fields. Just as in section 8 of Chapter 4, let us consider a geodesic $\gamma(t)$ for which

$$\gamma(0) = \bar{p}, \quad \bar{\gamma}(1) = \bar{q},$$

and let us now assume, in addition, that the point $\bar{q}$ belongs to some submanifold N and that

$$\dot{\gamma}(1) \perp N_{\bar{q}}.$$

Just as before, we define the space Θ merely by replacing the condition $X(1) = 0$ by the condition $X(1) \in N_{\bar{q}}$. We require now that δ be a positive number such that a δ-neighborhood of the point $\bar{q}$ on the submanifold N will be a coordinate neighborhood of some system of local coordinates that vanish at the point $\bar{q}$. We denote these coordinates by

$$z_{s+1}^1, \ \ldots, \ z_{s+1}^n \tag{5}$$

and we assume them fixed once and for all (in analogy with the coordinates $z_i^1, \ldots, z_i^{m-1}$ on the submanifolds P_i, where $i = 1, \ldots, s$).

For the set U, we now take the open set

$$U = U_1 \times \ldots \times U_s \times U_{s+1}$$

of the space $R^{s(m-1)+n}$, where $U_1, \ldots, U_s \subset R^{m-1}$ are the same sets as before and $U_{s+1} \subset R^n$ is the image of the δ-neighborhood of the point $\bar{q}$ in the manifold N under the coordinate homeomorphism corresponding to the coordinates (5).

Each point $z \in U$ will now define $s + 1$ points

$$q_1 \in P_1, \ldots, q_s \in P_s, q_{s+1} \in N.$$

Accordingly, we now take for the broken geodesic $u^z(t)$ the broken curve consisting of the minimizing geodesics connecting successively the points

$$p_0 = \gamma(\theta) = q_0, q_1, \ldots, q_s, q_{s+1}.$$

(Thus, the only difference from the construction in section 8 of Chapter 4 consists in the fact that the point q_{s+1} does not necessarily coincide with the fixed point $\bar{q}$.) Just as before, we choose the parameter t on the curve $u^z(t)$ proportional to the arc length and so that

$$u^z(\theta) = p_0, \; u^z(1) = q_{s+1}.$$

We define the functions $t_i(z)$, the variations φ_i^z, and the fields $X^z(t)$ just as in section 8 of Chapter 4. The conclusion that

$$X^z(t) \in \Theta$$

remains in force (it suffices to note that $X^z(1) = \dot{a}_{s+1}(0)$ and to replace the reference to the equation $\dot{a}_{s+1}(0) = 0$ in the proof of the equation $\dot{j}^z(0) = 0$ with a reference to the relation $\dot{a}_{s+1}(0) \perp \dot{\gamma}(1)$, which is valid by virtue of the fact that $\dot{a}_{s+1}(0) \in N_{\bar{q}}$.)

Obviously, the theorem on isomorphism in section 9 of Chapter 4 remains valid (with the space $R^{s(m-1)}$ replaced by the space $R^{s(m-1)+n}$ and, analogously, with the space $L = M_{p_1} + \ldots + M_{p_s}$ replaced by the space $L = M_{p_1} + \ldots + M_{p_s} + N_{\bar{q}}$). In the remark on the complete definition of the field $X(t)$ at the end of section 9 of Chapter 4, we must add an arbitrary vector $X_{s+1} \in N_{\bar{q}}$.

Morse's form $Q_\theta(z)$, for $z \in R^{s(m-1)+n}$, is now defined exactly as in section 10 of Chapter 4. However, instead of formula (1) of section 10 of Chapter 4, we now obtain the formula

$$Q_\theta(z) = \frac{1}{l}\left[\left(\frac{\nabla \dot{a}_{s+1}^z}{d\tau}(0), \dot{\gamma}(1)\right) + I_\theta^1(X^z(t))\right],$$

because now, in general, $\dfrac{\nabla \dot{a}_{s+1}^z}{d\tau}(0) \neq 0$.

But the curve $a_{s+1}^z(\tau)$ is now a curve on N for which

$$\dot{a}_{s+1}^z(0) = X^z(1),$$

and, therefore (cf. section 1),

$$\left(\frac{\nabla \dot{a}_{s+1}^z}{d\tau}(0), \dot{\gamma}(1)\right) = \pi_{\dot{\gamma}(1)}(X^z(1)) = (T_{\dot{\gamma}(1)} X^z(1), X^z(1)). \tag{6}$$

Thus,

$$Q_\theta(z) = \frac{1}{l}\bar{I}_\theta^1(X^z(t)),$$

where $\bar{I}_\theta^1$ is the functional $\bar{I}_a^b$, constructed for $a=\theta$, $b=1$ and the operator T: $M_{\bar{q}} \to M_{\bar{q}}$ (which coincides with the operator $T_{\dot{\gamma}(1)}$ on the space $N_{\bar{q}}$ and is equal to 0 on its orthogonal complement).

It follows immediately from this formula, just as from formula (1) of section 10 of Chapter 4, that

The form $Q_\theta(z)$ is positive-definite for θ sufficiently close to one.

Here, instead of inequality (2) of section 7 of Chapter 4, we now use inequality (4) [which we have not yet proven] because, in general, $X^z(1) \neq 0$).

Proof that the corank c_θ and the negative index of inertia h_θ of the form $Q_\theta(z)$ are independent of the points $t_1, \ldots, t_s$ is carried over to the present case without change except that we must replace the reference to inequality (2) of section 7 of Chapter 4 with a reference to inequality (4).

In proving the theorem on the index of a point (cf. section 11, Chapter 4), we take for the space Θ_0 the subspace (of the space Θ) consisting of smooth Jacobi fields $X(t) \in \Theta$ such that

$$\frac{\nabla X}{dt}(1) + T_{\dot{\gamma}(1)}X(1) \perp N_{\bar{q}}.$$

With this choice of subspace Θ_0 we have the equation

$$\Theta_0 = \Lambda_\gamma(\theta) \cap \Lambda_\gamma^N(1),$$

which is analogous to the equation $\Theta_0 = \Lambda_\gamma(\Theta) \cap \Lambda_\gamma(1)$ in section 10 of Chapter 4.

Furthermore, just as in section 11 of Chapter 4, by using formula (3) of section 7 of Chapter 4, we obtain

$$Q_\theta(z) = \frac{1}{l}\Big[\Big(\frac{\nabla X^z}{dt}(1) + T_{\dot{\gamma}(1)}X^z(1),\ X^z(1)\Big) + \\ + \sum_{i=1}^{s}\Big(\frac{\nabla X^z}{dt}(t_i - 0) - \frac{\nabla X^z}{dt}(t_i + 0),\ X^z(t_i)\Big)\Big],$$

and hence

$$Q_\theta(z) = \frac{1}{l}\left(\frac{\nabla X_\rho^{s+1}}{dt}(1) + T_{\dot{\gamma}(1)}X_\rho^{s+1}(1),\ X_\sigma^{s+1}(1)\right) z_{s+1}^\rho z_{s+1}^\sigma + \\ + \left[\frac{1}{l}\sum_{i=1}^{s}\left(\frac{\nabla X_{k_1}^{j_1}}{dt}(t_i - 0) - \frac{\nabla X_{k_1}^{j_1}}{dt}(t_i + 0),\ X_k^j(t)\right)\right] z_{j_1}^{k_1} z_j^k,$$

where ρ and σ are summed from 1 to n, where j and j_1 are summed from 1 to s, and where k and k_1 are summed from 1 to $m-1$. (Here, as in section 11 of Chapter 4, the $X_k^j(t)$, where $j = 1, \ldots, s+1$ and $k = 1, \ldots, m-1$ or n, are the fields $X^z(t)$ corresponding to the standard basis e_k^j of the space $R^{s(m-1)+n}$.)

It follows from this, just as in section 11 of Chapter 4, that the partial derivatives $L_k^j(z)$ of the form Q_θ with respect to the z_j^k are given by the formulas

$$L_k^j(z)=\begin{cases}\frac{2}{l}\left(\frac{\nabla X^z}{dt}(1)+T_{\dot\gamma(1)}X^z(1),\ X_k^{s+1}(1)\right), \\ \qquad\qquad \text{if } j=s+1,\ k=1,\ \ldots,\ n; \\ \frac{2}{l}\left[\sum_{i=1}^{s}\left(\frac{\nabla X^z}{dt}(t_i-0)-\frac{\nabla X^z}{dt}(t_i+0),\ X_k^j(t_i)\right)\right], \\ \qquad\qquad \text{if } j=1,\ \ldots,\ s;\ k=1,\ \ldots,\ m-1.\end{cases}$$

Just as in section 11 of Chapter 4, it follows immediately from these formulas that the field $X^z(t)$ is smooth for $z\in C_\theta$ (where C_θ is the maximum subspace of the space $R^{s(m-1)+n}$ on which the form $Q_\theta(z)$ vanishes) and that

$$\frac{\nabla X^z}{dt}(1)+T_{\dot\gamma(1)}X^z(1)\perp N_{\bar q}.$$

Consequently, if $z\in C_\theta$, then $X^z(t)\in\Theta_0$.

Conversely, if $X^z(t)\in\Theta_0$, then $L_k^j(z)=0$ for all j and k and, therefore, $z\in C_\theta$.

Thus, we have proved that

The index $\lambda_N(\theta)$ of the point θ is equal to the corank c_θ of the form $Q_\theta(z)$.

The corresponding theorem for the index of an interval is

The index $\lambda_N(0,1)$ of the interval (0, 1) is equal to the negative index of inertia h of the form $Q(z)=Q_0(z)$.

This theorem is proven virtually without any changes. More precisely, proof of the inequality

$$h\leqslant\lambda_N(0,1)$$

proceeds just does as the proof of the corresponding inequality (1) of section 12 of Chapter 4 but, in the proof of the opposite inequality

$$h\geqslant\lambda_N(0,1)$$

we assume that the fields $A_{j,k}(t)$, for $j=1,\ \ldots,\ r$ and $k=1,\ \ldots,\ \lambda_j$, are the basis fields of the space $\Lambda_\gamma(a_j)\cap\Lambda_\gamma^N(1)$ (and not the space $\Lambda_\gamma(a_j)\cap\Lambda_\gamma(1)$, as in section 12 of Chapter 4). We also replace the space $R^{s(m-1)}$ with the space $R^{s(m-1)+n}$. Finally, in proving the equation $\bar I_0^1(X(t))=0$ (which takes the place of the equation $I_0^1(X(t))=0$), we must remember that the extra term

$$\left(\frac{\nabla X}{dt}(1)+T_{\dot\gamma(1)}X(1),\ X(1)\right)$$

is equal to 0 since the field $X(t)$ coincides on the interval $[a_r,\ 1]$ with some field in $\Lambda_\gamma^N(1)$ and hence satisfies the relation

$$\frac{\nabla X}{dt}(1)+T_{\dot\gamma(1)}X(1)\perp N_{\bar q}.$$

Bott's form $Q_*(x)$ is now constructed just as in section 13 of Chapter 4 except that, for the set V, we now take the set

$$V = V_1 \times \dots \times V_s \times U_{s+1},$$

where $V_1, \dots, V_s \subset R^m$ are the same sets as in section 13 of Chapter 4 and $U_{s+1} \subset R^n$, just as above, is the image of the δ-neighborhood of the point $\bar{q}$ in the submanifold N under the corresponding coordinate homeomorphism.

The new set V is an open subset of the space R^{sm+n}, and every point x in it defines the $s + 1$ points

$$q_1 \in M, \dots, q_s \in M, q_{s+1} \in N.$$

The definition of the geodesic $u^x(t)$ is modified accordingly.

Since the modifications in the definition of the form $Q_*(x)$ are compatible with the modifications in the definition of the form $Q(z)$, we obtain (under the same assumptions as in section 13 of Chapter 4 and by repeating word for word the reasoning in that section) the result that the restriction $Q_*(z)$ of the form $Q_*(x)$ to the subspace $R^{s(m-1)+n}$ of the space R^{sm+n} defined by the equations

$$x_1^m = 0, \dots, x_s^m = 0,$$

differs from the form $Q(z)$ only by the factor $2l/(s + 1)$. (In the proof of the equation $\dot{J}_i(z) = 0$, we need, for $i = s$, to replace the reference to the equation $\dot{a}_{s+1}^z(0) = 0$ with a reference to the relation $\dot{a}_{s+1}^z(0) \perp \dot{\gamma}(1)$.)

Obviously, the new form $Q_*(x)$ coincides on the space R^s of the coordinates $x_1^m, \dots, x_s^m$ with the old form $Q_*(x)$. Therefore, this form is also positive-definite (on R^s).

Finally, in proving the formula

$$Q_*(x + z) = Q_*(x) + Q_*(z),$$

where z now belongs to $R^{s(m-1)+n}$ (whereas, before, $x \in R^s$), it suffices to replace the reference to the equation

$$\frac{\nabla \dot{a}_{s+1}^{x+z}}{d\tau}(0) = 0$$

with a reference to formula (6) and to recall that

$$\frac{\nabla \dot{a}_{s+1}^{x+z}}{d\tau} = \frac{\nabla \dot{a}_{s+1}^{z}}{d\tau}(0).$$

Otherwise, the reasoning in section 13 of Chapter 4 remains unchanged.

Summing up, we have

With the modifications indicated (for the most part, obvious), all the results (derived in section 4) regarding conjugate points and their indices remain valid for focal points and their indices.

Admittedly, this assertion is temporarily of a conditional nature since we still have not proven inequality (4).

4. Proof of inequality (4)

To prove inequality (4), let us find a bound from below for the functional

$$\overline{I}_a^b(X(t)) = (TX(b), X(b)) + \\ + \int_a^b \left[\left(\frac{\nabla X}{dt}(t), \frac{\nabla X}{dt}(t) \right) - R_{\gamma(t)}(X(t), \dot{\gamma}(t), \dot{\gamma}(t), X(t)) \right] dt$$

on the set $\mathfrak{M}_0$ of all fields $X(t) \in \mathfrak{M}_\gamma$ for which $X(a) = 0$.

Since every field $X(t) \in \mathfrak{M}_0$ is a covariant integral of the field $\frac{\nabla X}{dt}(t)$ that vanishes at the point a, we have, in accordance with formula (1) of section 5 of Chapter 4, the following equation for every $t \in [a, b]$ and every vector $B_t \in M_{\gamma(t)}$:

$$(B_t, X(t)) = \int_a^t \left(B_t, \tau_\theta^t \frac{\nabla X}{dt}(\theta) \right) d\theta, \tag{1}$$

where τ_θ^t: $M_{\gamma(\theta)} \to M_{\gamma(t)}$ is a parallel translation along the curve $\gamma(t)$. If we now set

$$B_t = R_{\gamma(t)}(Y(t), \dot{\gamma}(t)) \dot{\gamma}(t),$$

where $Y(t)$ is an arbitrary field in $\mathfrak{M}$, and remember that, by definition,

$$(R_{\gamma(t)}(Y(t), \dot{\gamma}(t)) \dot{\gamma}(t), X(t)) = R_{\gamma(t)}(Y(t), \dot{\gamma}(t), \dot{\gamma}(t), X(t)),$$

we immediately obtain the equation

$$R_{\gamma(t)}(Y(t), \dot{\gamma}(t), \dot{\gamma}(t), X(t)) = \\ = \int_a^t R_{\gamma(t)}\left(Y(t), \dot{\gamma}(t), \dot{\gamma}(t), \tau_\theta^t \frac{\nabla X}{dt}(\theta) \right) d\theta,$$

which is valid for an arbitrary field $Y(t) \in \mathfrak{M}$.

On the other hand, since

$$R_{\gamma(t)}\left(Y(t), \dot{\gamma}(t), \dot{\gamma}(t), \tau_\theta^t \frac{\nabla X}{dt}(\theta) \right) = \\ = \left(R_{\gamma(t)}\left(\tau_\theta^t \frac{\nabla X}{dt}(\theta), \dot{\gamma}(t) \right) \dot{\gamma}(t), Y(t) \right),$$

by virtue of the symmetry properties of a Riemannian tensor, if we now set $Y(t) = X(t)$ and again apply formula (1) (with $B_t = R_{\gamma(t)}\left(\tau_\theta^t \frac{\nabla X}{dt}(\theta),\ \dot\gamma(t)\right)\dot\gamma(t)$), we finally obtain

$$R_{\gamma(t)}(X(t),\ \dot\gamma(t),\ \dot\gamma(t),\ X(t)) = \\ = \int_a^t \int_a^t R_{\gamma(t)}\left(\tau_\theta^t \frac{\nabla X}{dt}(\theta),\ \dot\gamma(t),\ \dot\gamma(t),\ \tau_{\theta_1}^t \frac{\nabla X}{dt}(\theta_1)\right) d\theta\, d\theta_1.$$

For arbitrary $t \in [a, b]$, the expression

$$R_{\gamma(t)}\left(\tau_\theta^t \frac{\nabla X}{dt}(\theta),\ \dot\gamma(t),\ \dot\gamma(t),\ \tau_{\theta_1}^t \frac{\nabla X}{dt}(\theta_1)\right).$$

is a bilinear form in the vectors

$$\tau_\theta^t \frac{\nabla X}{dt}(\theta) \text{ and } \tau_{\theta_1}^t \frac{\nabla X}{dt}(\theta_1)$$

Therefore, according to a well-known theorem in linear algebra, it satisfies an inequality of the form

$$\left| R_{\gamma(t)}\left(\tau_\theta^t \frac{\nabla X}{dt}(\theta),\ \dot\gamma(t),\ \dot\gamma(t),\ \tau_{\theta_1}^t \frac{\nabla X}{dt}(\theta_1)\right)\right| \leqslant \\ \leqslant R(t)\left|\tau_\theta^t \frac{\nabla X}{dt}(\theta)\right|\left|\tau_{\theta_1}^t \frac{\nabla X}{dt}(\theta_1)\right|,$$

where $R(t)$ is some positive number. On the other hand, because of the isometry of a parallel translation, we have

$$\left|\tau_\theta^t \frac{\nabla X}{dt}(\theta)\right| = \left|\frac{\nabla X}{dt}(\theta)\right|$$

and

$$\left|\tau_{\theta_1}^t \frac{\nabla X}{dt}(\theta_1)\right| = \left|\frac{\nabla X}{dt}(\theta_1)\right|.$$

Therefore,

$$\left|R_{\gamma(t)}(X(t),\ \dot\gamma(t),\ \dot\gamma(t),\ X(t))\right| \leqslant \\ \leqslant \int_a^t \int_a^t \left| R_{\gamma(t)}\left(\tau_\theta^t \frac{\nabla X}{dt}(\theta),\ \dot\gamma(t),\ \dot\gamma(t),\ \tau_{\theta_1}^t \frac{\nabla X}{dt}(\theta_1)\right)\right| d\theta\, d\theta_1 \leqslant \\ \leqslant \int_a^b \int_a^b R(t)\left|\frac{\nabla X}{dt}(\theta)\right|\left|\frac{\nabla X}{dt}(\theta_1)\right| d\theta\, d\theta_1 \leqslant R(t)\left[\int_a^b \left|\frac{\nabla X}{dt}(\theta)\right| d\theta\right]^2$$

and, consequently,

$$\left|\int_a^b R_{\gamma(t)}(X(t),\ \dot\gamma(t),\ \dot\gamma(t),\ X(t))\, dt\right| \leqslant R_0\left[\int_a^b \left|\frac{\nabla X}{dt}(\theta)\right| d\theta\right]^2,$$

where

$$R_0 = \int_a^b R(t)\,dt.$$

Let us now find an analogous inequality for $(TX(b), X(b))$. If we apply formula (1) to the number $t = b$, we see that

$$(B, X(b)) = \int_a^b \left(B, \tau_\theta^b \frac{\nabla X}{dt}(\theta)\right) d\theta \tag{2}$$

for an arbitrary vector $B \in M_{\gamma(b)}$. In particular,

$$(TX(b), X(b)) = \int_a^b \left(TX(b), \tau_\theta^b \frac{\nabla X}{dt}(\theta)\right) d\theta.$$

But, since the operator T is self-adjoint,

$$\left(TX(b), \tau_\theta^b \frac{\nabla X}{dt}(\theta)\right) = \left(T\tau_\theta^b \frac{\nabla X}{dt}(\theta), X(b)\right).$$

Therefore, if we again apply formula (2) (with $B = T\tau_\theta^b \frac{\nabla X}{dt}(\theta)$), we obtain

$$(TX(b), X(b)) = \int_a^b \int_a^b \left(T\tau_\theta^b \frac{\nabla X}{dt}(\theta), \tau_{\theta_1}^b \frac{\nabla X}{dt}(\theta_1)\right) d\theta\, d\theta_1.$$

If we bound the bilinear form constituting the integrand in this expression just as we did above, we obtain the inequality

$$\left|\left(T\tau_\theta^b \frac{\nabla X}{dt}(\theta), \tau_{\theta_1}^b \frac{\nabla X}{dt}(\theta_1)\right)\right| \leqslant T_0 \left|\frac{\nabla X}{dt}(\theta)\right| \left|\frac{\nabla X}{dt}(\theta_1)\right|,$$

where T_0 is a positive number. Therefore,

$$|(TX(b), X(b))| \leqslant \int_a^b \int_a^b \left|\left(T\tau_\theta^b \frac{\nabla X}{dt}(\theta), \tau_{\theta_1}^b \frac{\nabla X}{dt}(\theta_1)\right)\right| d\theta\, d\theta_1 \leqslant$$

$$\leqslant \int_a^b \int_a^b T_0 \left|\frac{\nabla X}{dt}(\theta)\right| \left|\frac{\nabla X}{dt}(\theta_1)\right| d\theta\, d\theta_1 =$$

$$= T_0 \left[\int_a^b \left|\frac{\nabla X}{dt}(\theta)\right| d\theta\right]^2.$$

Note now, that, by virtue of the Cauchy-Schwarz-Bunyakovskiy integral inequality,

$$\left[\int_a^b \left|\frac{\nabla X}{dt}(\theta)\right| d\theta\right]^2 \leqslant \int_a^b \left|\frac{\nabla X}{dt}(\theta)\right|^2 d\theta \int_a^b 1 \cdot d\theta =$$

$$= (b-a)\int_a^b \left(\frac{\nabla X}{dt}(\theta), \frac{\nabla X}{dt}(\theta)\right) d\theta.$$

Consequently,

$$\left|\int_a^b R_{\gamma(t)}(X(t), \dot{\gamma}(t), \dot{\gamma}(t), X(t))\, dt\right| \leqslant$$

$$\leqslant R_0(b-a)\int_a^b \left(\frac{\nabla X}{dt}(t), \frac{\nabla X}{dt}(t)\right) dt,$$

$$|(TX(b), X(b))| \leqslant T_0(b-a)\int_a^b \left(\frac{\nabla X}{dt}(t), \frac{\nabla X}{dt}(t)\right) dt.$$

From these inequalities, we immediately get the following inequality for the functional $\bar{I}_a^b$:

$$\bar{I}_a^b(X(t)) \geqslant [1-(T_0+R_0)(b-a)]\int_a^b \left(\frac{\nabla X}{dt}(t), \frac{\nabla X}{dt}(t)\right) dt,$$

which is valid for an arbitrary field $X(t) \in \mathfrak{M}_0$.

Note now that, as a approaches b, the number T_0 does not change and the number R_0 can only decrease. Therefore, for a sufficiently close to b, the number

$$1-(T_0+R_0)(b-a)$$

is positive. Since the integral

$$\int_a^b \left(\frac{\nabla X}{dt}(t), \frac{\nabla X}{dt}(t)\right) dt$$

is obviously positive for an arbitrary not-identically-zero field $X(t) \in \mathfrak{M}_0$, this completes the proof of inequality (4) of section 3.

Remark 1: For $T = 0$, we obtain, in particular, a new proof of inequality (2) of section 7 of Chapter 4 if the points a and b are sufficiently close to each other. But when this inequality is established for sufficiently close a and b, we can, by using elementary continuity considerations, prove it without difficulty even for the case in which the interval $[a, b]$ contains no points conjugate with the point b. (This is true because if, for a point a_0, there exists a nonzero field $X(t) \in \mathfrak{M}$ for which $X(a_0) = 0$, $X(b) = 0$, and $I_{a_0}^b(X(t)) = 0$ and if inequality (2) of section 7 of Chapter 4 holds for an arbitrary point $a \in (a_0, b)$, then, as can easily be shown, the field $X(t)$ is a smooth Jacobi field and hence the point a_0 is conjugate with the point b.) Then, by using certain standard theorems from functional analysis, we can prove the fundamental inequality (1) of section 7 of Chapter 4. Thus, all the results of section 7 of Chapter 4 can be obtained without using Hilbert's integral. We suggest that the reader carry out the details of this reasoning.

Remark 2: The fundamental equation (3) of section 3 of Chapter 4 belongs to the class of so-called self-adjoint differential equations (Sturm-Liouville equations). It turns out that almost all the results (in particular, the theorems on the indices) in section 4 can be carried over without difficulty to all such equations. However, this would take us beyond the framework of our exposition.

CHAPTER 5

A REDUCTION THEOREM

1. Formulation of the theorem

Throughout this chapter, we shall assume that the Riemannian space M in question is *connected* and *compact*. As we know (cf. section 10 of Chapter 3), such a space is complete and hence all the results of the preceding chapter are applicable to it. Furthermore, by making an insignificant complication in our reasoning, we can replace the requirement of compactness with the other requirement of completeness.

Let $\bar{p}$ and $\bar{q}$ denote two distinct points in the space M and let $\hat{\Omega}$ denote the set of all piecewise-smooth curves connecting these two points. We shall assume that the parameter t on each curve $\gamma \in \hat{\Omega}$ is proportional to the arc length measured from the point $\bar{p}$ and that the coefficient of proportionality is equal to the reciprocal of the length $J(\gamma)$ of the curve γ. With this choice of parameter, we have $\gamma(0) = \bar{p}$ and $\gamma(1) = \bar{q}$ for an arbitrary curve $\gamma \in \hat{\Omega}$.

For each geodesic $\gamma \in \hat{\Omega}$, we shall denote the index $\lambda_1(0)$ of the point $t = 0$ with respect to the point $t = 1$ by $\lambda^0(\gamma)$ and we shall denote the index $\lambda_1(0, 1)$ of the interval $(0, 1)$ [again with respect to the point $t = 1$] by $\lambda^{0,1}(\gamma)$. We shall call the number $\lambda^0(\gamma)$ the corank and the number $\lambda^{0,1}(\gamma)$ the index of the geodesic γ.

Let us define a metric on the space $\hat{\Omega}$ by taking as the distance between two curves $\gamma_1, \gamma_2 \in \hat{\Omega}$ the number

$$\hat{\rho}(\gamma_1, \gamma_2) = \max_{0 \leqslant t \leqslant 1} [\rho(\gamma_1(t), \gamma_2(t))] + |J(\gamma_1) - J(\gamma_2)|. \tag{1}$$

Obviously, this distance is nondegenerate (that is, ($\hat{\rho}(\gamma_1, \gamma_2) = 0$ only for $\gamma_1 = \gamma_2$) and symmetric (that is, ($\hat{\rho}(\gamma, \gamma_2) = \rho(\gamma_2, \gamma_1)$). Let us show that the distance $\hat{\rho}(\gamma_1, \gamma_2)$ satisfies the triangle inequality, that is, that

$$\hat{\rho}(\gamma_1, \gamma_3) \leqslant \hat{\rho}(\gamma_1, \gamma_2) + \hat{\rho}(\gamma_2, \gamma_3)$$

for any three curves $\gamma_1, \gamma_2, \gamma_3 \in \hat{\Omega}$. Let t_0 denote a value of the parameter t such that

$$\rho(\gamma_1(t_0), \gamma_3(t_0)) = \max_{0 \leqslant t \leqslant 1} \rho(\gamma_1(t), \gamma_3(t)).$$

Then,

$$\begin{aligned}\hat{\rho}(\gamma_1, \gamma_3) = \rho(\gamma_1(t_0), \gamma_3(t_0)) + |J(\gamma_1) - J(\gamma_3)| \leqslant \\ \leqslant \rho(\gamma_1(t_0), \gamma_2(t_0)) + |J(\gamma_1) - J(\gamma_2)| + \\ + \rho(\gamma_2(t_0), \gamma_3(t_0)) + |J(\gamma_2) - J(\gamma_3)| \leqslant \\ \leqslant \hat{\rho}(\gamma_1, \gamma_2) + \hat{\rho}(\gamma_2, \gamma_3).\end{aligned}$$

Thus,

The distance (1) makes $\hat{\Omega}$ a metric space.

Obviously,

The function J is continuous on the metric space $\hat{\Omega}$.

(This is true because, by definition, $|J(\gamma_1) - J(\gamma_2)| \leqslant \hat{\rho}(\gamma_1, \gamma_2)$.)

Study of the properties of the space $\hat{\Omega}$ [and, in particular, study of the functions $\lambda^0(\gamma)$ and $\lambda^{0,1}(\gamma)$] is based on the remarkable analogy between this space and smooth manifolds. This analogy rests on the simple fact that the geodesics $\gamma \in \hat{\Omega}$ are stationary points of the function J (cf. section 1 of Chapter 4) and hence their properties must be similar to the properties of critical curves (that is, points at which $d\varphi = 0$ [cf. section 14, Chapter 1]) of smooth functions on smooth manifolds.

Let φ denote a smooth function on some smooth manifold M_* of dimension m_*. The value $d\varphi(X)$ of the differential of this function on some field $X \in \mathfrak{G}^1(M_*)$ is a smooth function on M_*. Therefore, we can speak of its differential $d[d\varphi(X)] \in \mathfrak{G}_1(M_*)$. For any two fields $X, Y \in \mathfrak{G}^1(M_*)$, we obtain

$$H\varphi(X, Y) = d[d\varphi(X)](Y) - \frac{1}{2} d\varphi([X, Y]).$$

Obviously, this function $H\varphi$ defined for all pairs of fields $X, Y \in \mathfrak{G}^1(M)$ is linear with respect to each argument; that is, it is a bilinear form on the linear space $\mathfrak{G}^1(M_*)$. (However, it is not $\mathfrak{G}(M_*)$-linear; that is, it is not a tensor field.) As one can easily see, the form $H\varphi$ is expressed in local coordinates by the formula

$$H\varphi(X, Y) = \frac{\partial^2\varphi}{\partial x^i \partial x^j} X^i Y^j + \frac{1}{2} \frac{\partial\varphi}{\partial x^j}\left(\frac{\partial Y^j}{\partial x^i} X^i + \frac{\partial X^j}{\partial x^i} Y^i\right), \quad (2)$$

from which it follows that

The form $H\varphi(X, Y)$ is symmetric.

We shall call the corresponding quadratic form

$$H\varphi(X) = H\varphi(X, X)$$

the Hessian of the function φ.

In accordance with formula (2), the value $H\varphi(X, Y)(p_*)$ of the form $H\varphi(X, Y)$ at a critical point p_* of the function φ is given by the formula

$$H\varphi(X, Y)(p_*) = \left(\frac{\partial^2\varphi}{\partial x^i \partial x^j}\right)_{p_*} X^i_{p_*} Y^j_{p_*} \quad (3)$$

Therefore, it depends only on the vectors X_{p_*} and Y_{p_*}. In other words, the formula

$$H_{p_*}\varphi(A, B) = H\varphi(X, Y)(p_*),$$

where $A, B \in (M_*)_{p_*}$ and X and Y are vector fields on M_* such that

$$A = X_{p_*}, \quad B = Y_{p_*},$$

defines unambiguously a symmetric bilinear form $H_{p_*}\varphi$ on the space $(M_*)_{p_*}$. We shall call the corresponding quadratic form

$$H_{p_*}\varphi(A) = H_{p_*}\varphi(A, A), \quad A \in (M_*)_{p_*}$$

the Hessian of the function φ at the point $p_* \in M_*$.

In accordance with formula (3), in an arbitrary system of local coordinates $x^1, \ldots, x^{m_*}$ at the point p_* we have

$$H_{p_*}\varphi(A, B) = \left(\frac{\partial^2\varphi}{\partial x^i\, \partial x^j}\right)_{p^*} A^i B^j, \quad A, B \in (M_*)_{p_*}$$

where the A^i and the B^j are the components of the vectors A and B relative to the basis

$$\left(\frac{\partial}{\partial x^1}\right)_{p_*}, \ldots, \left(\frac{\partial}{\partial x^{m_*}}\right)_{p_*}.$$

We emphasize that the form $H_{p_*}\varphi$ is defined only for *critical* p_* of the function φ.

We shall refer to the corank $c(p_*)$ and the negative index of inertia $h(p_*)$ of the quadratic form $H_{p_*}\varphi$ as the corank and index of the critical point p_* respectively. As we shall see below, these numbers are analogous to the numbers $\lambda^0(\gamma)$ and $\lambda^{0,1}(\gamma)$ respectively.

From heuristic considerations let us turn to precise mathematical formulations. Let us consider the subspaces $[J \leqslant a]$ (obviously closed) of the space $\hat{\Omega}$ consisting of all curves $\gamma \in \hat{\Omega}$ for which $J(\gamma) \leqslant a$ and, analogously, the subspaces $[\varphi \leqslant b]$ of the smooth manifold M_* that consists of all points $p_* \in M_*$ for which $\varphi(p_*) \leqslant b$. Here, a and b are arbitrary real numbers. (For these subspaces to be nonempty, we must choose the numbers a and b sufficiently close to each other; for example, the subspace $[J \leqslant a]$ will be nonempty if and only if $a \geqslant \rho(\bar{p}, \bar{q})$.)

The purpose of the present chapter is to prove the following reduction theorem of Bott, which reduces the problem of studying the subspace $[J \leqslant a]$ to the problem of studying the subspaces $[\varphi \leqslant b]$:

For an arbitrary number $a \geqslant \rho(\bar{p}, \bar{q})$, there exists a smooth manifold M_*, a smooth function φ on M_*, and a number b such that the space $[J \leqslant a]$ is of the same homotopic type as the space $[\varphi \leqslant b]$. Furthermore, there exists a homotopic equivalence

$$\alpha\colon\ [J \leqslant a] \to [\varphi \leqslant b],$$

such that the curve $\gamma \in [J \leqslant a]$ *will be a geodesic if and only if the point* $\alpha(\gamma) \in [\varphi \leqslant b]$ *is a critical point of the function* φ. *Finally,*

(1) *The corank* $c(\alpha(\gamma))$ *of the point* $\alpha(\gamma)$ *is equal to the corank* $\lambda^0(\gamma)$ *of the geodesic* γ;

(2) *the index* $h(\alpha(\gamma))$ *of the point* $\alpha(\gamma)$ *is equal to the index* $\lambda^{0,1}(\gamma)$ *of the geodesic* γ.

(We recall that topological spaces X and Y are said to be of the same homotopic type if there exist continuous mappings

$$\alpha: X \to Y, \quad \beta: Y \to X,$$

such that the composite mappings

$$\beta \circ \alpha: X \to X, \quad \alpha \circ \beta: Y \to Y$$

are homotopic to the identity mappings 1_X and 1_Y of the spaces X and Y respectively, that is, if there exist families of continuous mappings

$$D_\tau: X \to X, \quad \Delta_\tau: Y \to Y,$$

depending continuously on the parameter $\tau \in [0, 1]$ such that

$$D_0 = \beta \circ \alpha, \quad D_1 = 1_X,$$
$$\Delta_0 = \alpha \circ \beta, \quad \Delta_1 = 1_Y.$$

Mappings α and β with these properties are called (mutually inverse) *homotopic equivalences*, and the families D_τ and Δ_τ are called *deformations*.)

2. Remarks on the formulation of the reduction theorem

Since the space M is compact, there exists a positive number $d = d_M$ such that any two points $p, q \in M$ at a distance $\rho(p, q) < d$ less than d from each other can be joined by a unique minimizing curve (cf. section 7, Chapter 3). The points p and q on this minimizing curve are not conjugate to each other (cf. section 4, Chapter 4). Let us assume that such a number d is chosen once and for all.

We shall prove the reduction theorem by taking for the manifold M_* the Cartesian product

$$M_* = \underbrace{M \times \ldots \times M}_{s \text{ times}}$$

(cf. section 15, Chapter 1), where s is an integer so large that

$$\frac{a}{\sqrt{s+1}} < d,$$

and by defining the function φ and the number b by

$$\varphi(p_*)=\rho^2(\bar{p},\ p_1)+\ \dots\ +\rho^2(p_i,\ p_{i+1})+\ \dots\ +\rho^2(p_s,\ \bar{q}),$$
$$p_*=(p_1,\ \dots,\ p_s)\in M_*,\quad p_1,\ \dots,\ p_s\in M, \tag{1}$$
$$b=\frac{a^2}{s+1}.$$

We define the mapping

$$\alpha:\ [J\leqslant a]\to[\varphi\leqslant b]$$

by

$$\alpha(\gamma)=\left(\gamma\left(\frac{1}{s+1}\right),\ \dots,\ \gamma\left(\frac{i}{s+1}\right),\ \dots,\ \gamma\left(\frac{s}{s+1}\right)\right).$$

For the mapping β (which, as we shall show, will be the homotopic equivalence inverse to the mapping α), we take the mapping

$$\beta:\ [\varphi\leqslant b]\to[J\leqslant a],$$

that assigns to each point $p_*=(p_1,\ \dots,\ p_s)\in[\varphi\leqslant b]$ the broken geodesic $\beta(p_*)$ composed of the minimizing curves connecting successively the points p_i and p_{i+1}, where $i=0,1,\dots,s$. (We assume that $p_0=\bar{p}$ and $p_{s+1}=\bar{q}$).)

Obviously,

The function φ is a smooth function on the manifold M_*.

(This follows immediately from the smoothness of the function ρ^2 [cf. section 9, Chapter 3].)

The mapping α maps the space $[J\leqslant a]$ into the space $[\varphi\leqslant b]$

$$\left(\text{since } \sum_{i=0}^{s}\rho^2\left(\gamma\left(\frac{i}{s+1}\right),\ \gamma\left(\frac{i+1}{s+1}\right)\right)=\sum_{i=0}^{b}\left(\frac{J(\gamma)}{s+1}\right)^2\leqslant\frac{a^2}{s+1}=b\right)$$

The mapping β is uniquely defined.

(This is true because $\rho^2(p,\ p_{i+1})\leqslant\varphi(p_*)\leqslant b$ and hence $\rho(p_i,\ p_{i+1})<d$, so that each of the minimizing curves constituting the broken geodesic $\beta(p_*)$ is uniquely defined.)

The mapping β maps the space $[\varphi\leqslant b]$ into the space $[J\leqslant a]$.

This is true because[1]

$$J^2(\beta(p_*))=\left(\sum_{i=0}^{s}\rho(p_i,\ p_{i+1})\right)^2\leqslant(s+1)\sum_{i=0}^{s}\rho^2(p_i,\ p_{i+1})\leqslant(s+1)b=a^2).$$

Obviously, the points

$$t_1=\frac{1}{s+1},\ \dots,\ t_s=\frac{s}{s+1},$$

---

*Here, we use a special case of the Cauchy-Schwarz-Bunyakovskiy inequality

$$(a_1+\ \dots\ +a_n)^2\leqslant n(a_1^2+\ \dots\ +a_n^2).$$

are distributed so densely (in the sense of section 8, Chapter 4) on the interval [0, 1] that they can be used for the points $t_1, \ldots, t_s$ in the basis for the construction of Bott's form $Q_*(x)$. When we do this, we immediately obtain the result that, for an arbitrary geodesic $\gamma \in [J \leqslant a]$, the function φ in a neighborhood of the point $\alpha(\gamma) = (\gamma(t_1), \ldots, \gamma(t_s))$ is simply the function $J_*(x)$ considered in section 13 of Chapter 4; more precisely, the coordinate homeomorphism defined by the local coordinates $\widehat{x}_i^k$, for $i = 1, \ldots, s$ and $k = 1, \ldots, m$, at the point $\alpha(\gamma)$ of the manifold M_* that correspond to the local coordinates $\overset{i}{x_k}$ (see section 15 of Chapter 1) chosen in section 13 of Chapter 4 maps the function φ into the function J_*. This means, first of all, that the point $\alpha(\gamma)$ is a critical point of the function φ [cf. formula (6), section 13, Chapter 4] and, second, that the Hessian $H_{\alpha(\gamma)}\varphi$ of the function φ at the point $\alpha(\gamma)$ coincides with Bott's form $Q_*(x)$ [after appropriately identifying the space $(M_*)_{p_*}$ with the space R^{sm}]. Thus, we have proven two things: First,

For an arbitrary geodesic $\gamma \in [J \leqslant a]$, the point $\alpha(\gamma)$ is a critical point of the function φ.

Second (cf. the theorems on indices, Chapter 4),

Assertions (1) and (2) of the reduction theorem are valid.

Let us now show that

For an arbitrary critical point $p_* \in [\varphi \leqslant b]$ of the φ function, there is a geodesic $\gamma \in [J \leqslant a]$ such that

$$\alpha(\gamma) = p_*.$$

Proof: Let us choose, at the point $p_* = (p_1, \ldots, p_s)$ of the manifold M_*, an arbitrary vector $C = (C_1, \ldots, C_s)$, where

$$C_1 \in M_{p_1}, \ldots, C_s \in M_{p_s}.$$

Consider the square $\rho^2 = \rho^2(p, q)$ of the metric ρ as a smooth function on $M \times M$. In view of formula (1), we immediately obtain

$$C\varphi = \sum_{i=0}^{s} C_{(i)}\rho^2,$$

where $C_{(i)}$ is the vector (C_i, C_{i+1}) at the point (p_i, p_{i+1}) of the manifold $M \times M$ (we recall that $C_0 = 0$ and $C_{s+1} = 0$).

If $p_{i+1} \neq p_i$, then

$$C_{(i)}\rho^2 = 2\rho(p_i, p_{i+1})C_{(i)}\rho.$$

On the other hand, in accordance with formula (5) of section 1 of Chapter 4 (the conditions under which this formula is applicable, namely, sufficiently small distances between the points p_i and p_{i+1}, are obviously satisfied), we have

$$C_{(i)}\rho = (C_{i+1}, \dot{\gamma}_i^+) - (C_i, \dot{\gamma}_i^-),$$

where $\dot{\gamma}_i^-$ and $\dot{\gamma}_i^-$ are unit tangent vectors of the geodesic γ_i at the points p_i and p_{i+1} respectively. Consequently,

$$C_{(i)}\rho^2 = 2\rho(p_i, p_{i+1})[(C_{i+1}, \dot{\gamma}_i^+) - (C_i, \dot{\gamma}_i^-)].$$

Obviously, this formula is valid for $p_{i+1}=p_i$ (with both sides equal to zero). Therefore,

$$C\varphi = 2\sum_{i=0}^{s}\rho(p_i, p_{i+1})[(C_{i+1}, \dot{\gamma}_i^+) - (C_i, \dot{\gamma}_i^-)] = $$
$$= 2\sum_{i=1}^{s-1}(\rho(p_{i-1}, p_i)\dot{\gamma}_{i-1}^+ - \rho(p_i, p_{i+1})\dot{\gamma}_i^-, C_i),$$

since, by hypothesis, $C_0 = 0$ and $C_{s+1} = 0$.

But since, by hypothesis, the point p_* is a critical point of the function φ, we have $C\varphi = 0$ for an arbitrary vector $C \in (M_*)_{p_*}$. Consequently, for arbitrary vectors $C_i \in M_{p_i}$, where $i = 1, \ldots, s$, we have

$$\sum_{i=1}^{s-1}(\rho(p_{i-1}, p_i)\dot{\gamma}_{i-1}^+ - \rho(p_i, p_{i+1})\dot{\gamma}_i^-, C_i) = 0,$$

which is possible only when

$$\rho(p_{i-1}, p_i)\dot{\gamma}_{i-1}^+ = \rho(p_i, p_{i+1})\dot{\gamma}_i^-$$

for arbitrary $i = 1, \ldots, s$.

Since the vectors $\dot{\gamma}_i^-$ and $\dot{\gamma}_{i-1}^+$ are unit vectors, it follows, first of all, from this equation that

$$\rho(p_{i-1}, p_i) = \rho(p_i, p_{i+1}),$$

and then that

$$\dot{\gamma}_{i-1}^+ = \dot{\gamma}_i^-$$

for arbitrary $i = 1, \ldots, s$. (We note that this implies, in particular, that the equation $\rho(p_{i-1}, p_i) = 0$ is impossible for every $i = 1, \ldots, s$, since satisfaction of it for one i implies satisfaction for all i and the distance between the points $\bar{p}$ and $\bar{q}$ would, in contradiction with the hypothesis, be equal to 0.)

The equations we have obtained mean that the broken geodesic γ composed of the minimizing curves γ_i is indeed a smooth geodesic and that the mapping α maps this geodesic into the point $p_* = (p_1, \ldots, p_s)$. This completes the proof of the assertion made above.

When we consider all that has been said, we see that, to prove the reduction theorem, it only remains for us to show that

The mappings α and β are continuous and are inverse homotopic equivalences.

3. Continuity of the mappings α and β

To prove that the mappings α and β are continuous, it will be convenient to consider the space Φ of all piecewise-smooth curves in the space M (more precisely, all finite segments of them) that are parametrized by numbers in the interval $[0, 1]$ proportionate (except in the degenerate case) to arc length. In other words, points $\gamma=\gamma(t)$ in the space Φ are the continuous mappings

$$\gamma:[0,\ 1]\to M,$$

enjoying the properties that, first, these mappings are smooth in t for all except a finite number of points $t\in[0,\ 1]$ and, second, the vectors $\dot{\gamma}(t)=d\gamma\left(\frac{\partial}{\partial t}\right)$ have the same length (depending on γ) for all t (for which they are defined). Thus, every curve $\gamma\in\Phi$ is either piecewise-regular (and then has a nonzero length) or degenerate (and then has zero length). The space $\hat{\Omega}$ introduced in section 1 is clearly a subspace of the space Φ. (The curve $\gamma\in\Phi$ belongs to $\hat{\Omega}$ if and only if $\gamma(0)=\bar{p}$ and $\gamma(1)=\bar{q}$. Since $\bar{p}\neq\bar{q}$, all the curves $\gamma\in\hat{\Omega}$ are nondegenerate.)

Let us define a metric $\hat{\rho}$ in the space Φ by the same formula [(1) of section 1] as was used to define the metric in the space $\hat{\Omega}$. That this metric is nondegenerate and symmetric and that it satisfies the triangle inequality are proven in just the same way as the corresponding assertions were proven for the space $\hat{\Omega}$. Therefore,

The space Φ is a metric space containing the space $\hat{\Omega}$ as a subspace.

Now, let us show that

For an arbitrary curve $\gamma\in\Phi$ and an arbitrary $t\in[0,\ 1]$, the point $\gamma(t)\in M$ depends continuously on γ and t.

Proof: Since the point $\gamma(t)$ depends continuously on t, it follows that, for arbitrary positive ε, there exists a positive δ such that $\rho(\gamma(t),\ \gamma(t_1))<\varepsilon/2$ whenever $|t-t_1|<\delta$.

Let γ_1 denote an arbitrary curve in Φ such that $\hat{\rho}(\gamma,\ \gamma_1)<\varepsilon/2$. Then,

$$\rho(\gamma(t),\ \gamma_1(t_1))\leqslant\rho(\gamma(t),\ \gamma(t_1))+\rho(\gamma(t_1),\ \gamma_1(t_1))\leqslant\frac{\varepsilon}{2}+\hat{\rho}(\gamma,\ \gamma_1)<\varepsilon.$$

Thus, for arbitrary positive ε, there exist numbers $\delta>0$ and $\delta_1=\varepsilon/2>0$ such that $\rho(\gamma(t),\ \gamma_1(t_1))<\varepsilon$, whenever $|t-t_1|<\delta$, and $\rho(\gamma,\ \gamma_1)<\delta_1$. But this means that the point $\gamma(t)$ depends continuously on γ and t.

Since, by definition,

$$\alpha(\gamma)=\left\{\gamma\left(\frac{1}{s+1}\right),\ \ldots,\ \gamma\left(\frac{s}{s+1}\right)\right\}$$

it follows immediately from the assertion that we have just proved that

The mapping α is continuous.

Before proving that the mapping β is continuous, let us show that

For any two points $p,\ q\in M$ such that $\rho(p,\ q)<d$, the minimizing geodesic $\gamma_{p,q}$ connecting them depends continuously (since it is a point of the space Φ) on the points p and q.

Proof: In accordance with the theorem on the dependence of solutions of a system of differential equations on the initial conditions, the local coordinates of the point $\gamma_{p,q}(t)$ depend continuously (and, in fact, smoothly) on the point p. Therefore, the point $\gamma_{p,q}(t)$ also depends continuously on the point p and, hence, for arbitrary positive ε, there exists a positive δ such that the inequality $\rho(p, p_1) < \delta$ implies the inequality

$$\rho(\gamma_{p,q}(t), \gamma_{p_1,q}(t)) < \frac{\varepsilon}{4}.$$

Since the interval [0, 1] is compact, a single δ can be chosen for all $t \in [0, 1]$. Consequently, for $\rho(p, p_1) < \delta$, we have

$$\max_{0 \leqslant t \leqslant 1} \rho(\gamma_{p,q}(t), \gamma_{p_1,q}(t)) < \frac{\varepsilon}{4}.$$

On the other hand, the distance $\rho(p, q) = J(\gamma_{p,q})$ is, as we have seen, a continuous function (and a smooth function for $p \neq q$) of the points p and q. Therefore, for sufficiently small positive δ,

$$|J(\gamma_{p,q}) - J(\gamma_{p_1,q})| < \frac{\varepsilon}{4}.$$

Thus, if δ is a sufficiently small positive number, we have

$$\hat{\rho}(\gamma_{p,q}, \gamma_{p_1,q}) < \frac{\varepsilon}{2},$$

whenever $\rho(p, p_1) < \delta$. Analogously, the inequality $\rho(q, q_1) < \delta$ implies the inequality $\hat{\rho}(\gamma_{p_1,q}, \gamma_{p_1,q_1}) < \varepsilon/2$. Therefore,

$$\hat{\rho}(\gamma_{p,q}, \gamma_{p_1,q_1}) < \varepsilon,$$

whenever $\rho(p, p_1) < \delta$ and $\rho(q, q_1) < \varepsilon$. But this means that the curve $\gamma_{p,q}$ depends continuously on the points p and q.

Now, let γ_1 and γ_2 denote two curves in Φ such that

$$\gamma_1(1) = \gamma_2(0).$$

Then, we can define a new curve $\gamma \in \Phi$ by "attaching" the curve γ_1 to the curve γ_2 and transforming the parameter accordingly. We shall call this curve γ the ***composition*** of the curves γ_1 and γ_2 and we shall denote it by $\gamma_1 * \gamma_2$. If at least one of the curves γ_1, γ_2 is nondegenerate, then the curve $\gamma = \gamma_1 * \gamma_2$ is of length $J(\gamma) = J(\gamma_1) + J(\gamma_2) \neq 0$ and, as can easily be seen, is given by the formula

$$\gamma(t) = \begin{cases} \gamma_1\left(\dfrac{J(\gamma)}{J(\gamma_1)} t\right) & \text{if } 0 \leqslant t \leqslant \dfrac{J(\gamma_1)}{J(\gamma)}; \\ \gamma_2\left(\dfrac{J(\gamma)t - J(\gamma_1)}{J(\gamma_2)}\right) & \text{if } \dfrac{J(\gamma_1)}{J(\gamma)} \leqslant t \leqslant 1. \end{cases}$$

Obviously, if either γ_1 or γ_2 is degenerate, the composition

$\gamma_1 * \gamma_2$ coincides with the other curve. This remains true even when both γ_1 and γ_2 are degenerate (in which case, the composition $\gamma_1 * \gamma_2$ is also degenerate).

Together with the curves γ_1 and γ_2, let us now consider two other curves γ_1' and γ_2' for which the composition $\gamma_1' * \gamma_2'$ is also defined (that is, that enjoy the property that $\gamma_1'(1)=\gamma_2' 0)$. Suppose that

$$\hat{\rho}(\gamma_1, \gamma_1') < \varepsilon, \quad \hat{\rho}(\gamma_2, \gamma_2') < \varepsilon,$$

where ε is some positive number. Let us show that, for the distance $\hat{\rho}(\gamma, \gamma')$ between the curves $\gamma = \gamma_1 * \gamma_2$ and $\gamma' = \gamma_1' * \gamma_2'$, we have the inequality

$$\hat{\rho}(\gamma, \gamma') < 5\varepsilon. \tag{1}$$

Since

$$|J(\gamma) - J(\gamma')| = |J(\gamma_1) + J(\gamma_2) - J(\gamma_1') - J(\gamma_2')| \leqslant$$
$$\leqslant |J(\gamma_1) - J(\gamma_1')| + |J(\gamma_2) - J(\gamma_2')| < \varepsilon + \varepsilon = 2\varepsilon,$$

to prove inequality (1), it suffices to show that

$$\max_{0 \leqslant t \leqslant 1} \rho(\gamma(t), \gamma'(t)) < 3\varepsilon. \tag{2}$$

Suppose first that the curve γ is degenerate; that is, suppose that $\gamma(t) = p$ for all t, where p is some point in the space M. Then, for all t, we have the inequalities $\rho(p, \gamma_1'(t)) < \varepsilon$ and $\rho(p, \gamma_2'(t)) < \varepsilon$ and hence the inequality $\rho(p, \gamma'(t)) < \varepsilon$. Thus, in this case, we have

$$\max_{0 \leqslant t \leqslant 1} \rho(\gamma(t), \gamma'(t)) < \varepsilon.$$

Suppose now that the curve γ is nondegenerate and hence

$$J(\gamma) = J(\gamma_1) + J(\gamma_2) \neq 0.$$

Let us show that

If $J(\gamma_1) J(\gamma') = J(\gamma_1) J(\gamma)$, then

$$\max_{0 \leqslant t \leqslant 1} \rho(\gamma(t), \gamma'(t)) < \varepsilon.$$

If $0 \leqslant t \leqslant \frac{J(\gamma_1)}{J(\gamma)}$, we have

$$\gamma(t) = \gamma_1\left(\frac{J(\gamma)}{J(\gamma_1)} t\right), \quad \gamma'(t) = \gamma_1'\left(\frac{J(\gamma)}{J(\gamma_1)} t\right)$$

and hence

$$\rho(\gamma(t), \gamma'(t)) = \rho\left(\gamma_1\left(\frac{J(\gamma)}{J(\gamma_1)} t\right), \quad \gamma_1'\left(\frac{J(\gamma)}{J(\gamma_1)} t\right)\right) < \varepsilon.$$

Analogously, if $\frac{J(\gamma_1)}{J(\gamma)} \leqslant t \leqslant 1$, we have

$$\gamma(t) = \gamma_2\left(\frac{J(\gamma)t - J(\gamma_1)}{J(\gamma_2)}\right), \quad \gamma'(t) = \gamma_2'\left(\frac{J(\gamma)t - J(\gamma_1)}{J(\gamma_2)}\right),$$

and hence we again have

$$\rho(\gamma(t), \gamma'(t)) < \varepsilon.$$

In the general case, let us consider the segments $\gamma'|_{[0, J(\gamma_1)/J(\gamma)]}$ and $\gamma'|_{[J(\gamma_1)/J(\gamma), 1]}$ of the curve γ'. If we parametrize these segments suitably, we obtain two curves $\gamma^{(1)}$, $\gamma^{(2)} \in \Phi$ enjoying the properties that, first, $\gamma' = \gamma^{(1)} * \gamma^{(2)}$ and, second, $J(\gamma_1)J(\gamma') = J(\gamma^{(1)})J(\gamma)$. Let us find a bound for the distance $\hat{\rho}(\gamma_1', \gamma^{(1)})$.

Since the parameter t is proportional on all these curves (when they are nondegenerate) to the arc length, it follows that for arbitrary t, the arc length of the curve γ' between the points $\gamma^{(1)}(t)$ and $\gamma_1'(t)$ is equal to $t\,|J(\gamma^{(1)}) - J(\gamma_1')|$. But

$$|J(\gamma^{(1)}) - J(\gamma_1')| = \left|J(\gamma')\frac{J(\gamma_1)}{J(\gamma)} - J(\gamma_1')\right| =$$
$$= \left|\frac{J(\gamma_1)[J(\gamma_2') - J(\gamma_2)] + J(\gamma_2)[J(\gamma_1) - J(\gamma_1')]}{J(\gamma)}\right| <$$
$$< \frac{J(\gamma_1)\varepsilon + J(\gamma_2)\varepsilon}{J(\gamma)} = \varepsilon.$$

Therefore, this length is less than εt, that is, less than ε. Consequently, the distance $\rho(\gamma^{(1)}(t), \gamma_1'(t))$ is also less than ε. Since, as we have just shown, $|J(\gamma^{(1)}) - J(\gamma_1')| < \varepsilon$, this proves that

$$\hat{\rho}(\gamma_1', \gamma^{(1)}) < 2\varepsilon.$$

Consequently,

$$\hat{\rho}(\gamma_1, \gamma^{(1)}) \leqslant \hat{\rho}(\gamma_1, \gamma_1') + \hat{\rho}(\gamma_1', \gamma^{(1)}) < \varepsilon + 2\varepsilon = 3\varepsilon.$$

Analogously, we can show that

$$\hat{\rho}(\gamma_2, \gamma^{(2)}) < 3\varepsilon.$$

Therefore, in accordance with the assertion proven above (as applied to the curves $\gamma^{(1)}$ and $\gamma^{(2)}$ and to the number 3ε), we have

$$\max_{1 \leqslant t \leqslant 1} \rho(\gamma(t), \gamma'(t)) < 3\varepsilon.$$

This completes the proof of inequality (2) and hence of inequality (1).

Since $5\varepsilon \to 0$ as $\varepsilon \to 0$, it follows immediately from inequality (1) that

The curve $\gamma_1 * \gamma_2$ *depends continuously on the curves* γ_1 *and* γ_2.

It will now be easy for us to prove that the mapping β is continuous.

Let $p_* = (p_1, \ldots, p_s)$ denote an arbitrary point in the space $[\varphi \leqslant b]$. By definition, the curve $\beta(p_*)$ is the composition

$$\gamma_{\bar{p}, p_1} * \gamma_{p_1, p_2} * \ldots * \gamma_{p_s, \bar{q}} \tag{3}$$

of minimizing curves $\gamma_{p_i, p_{i+1}}$ connecting the points p_i and p_{i+1}, where $i = 1, \ldots, s$. (Just as above, we assume that $p_0 = \bar{p}$ and $p_{s+1} = \bar{q}$.) We showed above that every minimizing curve $\gamma_{p_i, p_{i+1}}$ depends continuously on the corresponding points p_i and p_{i+1}. On the other hand, according to what we have just proven, the composition (3) depends continuously on the minimizing curves $\gamma_{p_i, p_{i+1}}$. Consequently, the curve $\beta(p_*)$ depends continuously on the point p_*; that is,

The mapping β is continuous.

4. Completion of the proof of the reduction theorem

By definition, the mapping

$$\beta \circ \alpha : [J \leqslant a] \to [J \leqslant a]$$

assigns to every curve $\gamma \in [J \leqslant a]$ the broken geodesic with vertices at the points $\gamma\left(\frac{i}{s+1}\right)$, for $i = 0, \ldots, s+1$. To show that this mapping is homotopically identical to the mapping $1 = 1_{[J \leqslant a]}$ of the space $[J \leqslant a]$, we must construct a continuous family (deformation) of continuous mappings

$$D_\tau : [J \leqslant a] \to [J \leqslant a], \ 0 \leqslant \tau \leqslant 1,$$

such that

$$D_0 = 1, \ D_1 = \beta \circ \alpha$$

(cf. end of section 1).

Let us define the deformation D_τ by assigning to an arbitrary curve $\gamma \in [J \leqslant a]$ and an arbitrary number $\tau \in [0, 1]$ the curve

$$D_\tau(\gamma) = \bar{\gamma}_0^{(\tau)} * \gamma_0^{(\tau)} * \bar{\gamma}_1^{(\tau)} * \gamma_1^{(\tau)} * \ldots * \bar{\gamma}_s^{(\tau)} * \gamma_s^{(\tau)},$$

where $\bar{\gamma}_i^{(\tau)}$, for $i = 0, \ldots, s$, is the minimizing geodesic connecting the points

$$\gamma\left(\frac{i}{s+1}\right) \text{ and } \gamma\left(\frac{\tau+i}{s+1}\right)$$

(obviously, the distance between these points is less than d and hence the geodesic "chord" $\bar{\gamma}_i^{(\tau)}$ is unambiguously determined) and where $\gamma_i^{(\tau)}$ is the curve in Φ that is the parametrized segment

$$\gamma\Big|_{\left[\frac{\tau+i}{s+1},\ \frac{i+1}{s+1}\right]}$$

of the curve γ.

Obviously, $D_\tau(\gamma) \in [J \leqslant a]$ and

$$D_0(\gamma) = \gamma, \quad D_1(\gamma) = (\beta \circ \alpha)\gamma$$

for an arbitrary curve $\gamma \in [J \leqslant a]$. Therefore, it suffices to show that

The curve $D_\tau(\gamma)$ *depends continuously on the curve* γ *and the number* τ.

According to the results of the preceding section, the points $\gamma\left(\frac{i}{s+1}\right)$ and $\gamma\left(\frac{\tau+i}{s+1}\right)$ and hence the minimizing curve $\bar{\gamma}_i^{(\tau)}$ depend continuously on γ and τ. Furthermore, the curve $D_\tau(\gamma)$ depends continuously on the curves $\bar{\gamma}_i^{(\tau)}$ and $\gamma_i^{(\tau)}$. Therefore, it suffices to show that

The curve $\gamma_i^{(\tau)}$ *depends continuously on the curve* γ *and the number* τ.

Let us prove the following more general assertion:

For an arbitrary curve $\gamma \in \Phi$ *and arbitrary numbers* τ *and* θ *that satisfy the inequalities.*

$$0 \leqslant \tau \leqslant \theta \leqslant 1,$$

the curve $\gamma_\tau^\theta \in \Phi$ *is the parametrized segment* $\gamma|_{[\tau,\theta]}$ *of the curve* γ *and it depends continuously on the curve* γ *and the numbers* τ *and* θ.

(For $\tau = \theta$, the curve γ_τ^θ is understood to mean the corresponding degenerate curve.)

Proof: To prove this assertion, let us consider an arbitrary curve $\gamma' \in \Phi$ and arbitrary numbers τ' and θ' (satisfying the inequalities $0 \leqslant \tau' \leqslant \theta' \leqslant 1$) such that

$$\hat{\rho}(\gamma, \gamma') < \varepsilon, \quad |\tau - \tau'| < \varepsilon, \quad |\theta - \theta'| < \varepsilon,$$

where ε is some positive number. To find a bound for the distance $\hat{\rho}(\gamma_\tau^\theta, \gamma'^{\theta'}_{\tau'})$, let us consider the auxiliary curve $\gamma_{\tau'}^{\theta'}$. We note, first of all, that the points on the curves $\gamma_{\tau'}^{\theta'}$ and $\gamma'^{\theta'}_{\tau'}$, corresponding to a single value of the parameter will also enjoy this property when we treat them as points on the curves γ and γ' respectively. Therefore,

$$\max_{0 \leqslant t \leqslant 1} \rho\left(\gamma_{\tau'}^{\theta'}(t), \gamma'^{\theta'}_{\tau'}(t)\right) \leqslant \max_{0 \leqslant t \leqslant 1} \rho(\gamma(t), \gamma'(t)).$$

Also,

$$|J(\gamma_{\tau'}^{\theta'}) - J(\gamma'^{\theta'}_{\tau'})| = (\theta' - \tau')|J(\gamma) - J(\gamma')| \leqslant |J(\gamma) - J(\gamma')|.$$

Consequently,

$$\hat{\rho}(\gamma_{\tau'}^{\theta'}, \gamma'^{\theta'}_{\tau'}) \leqslant \hat{\rho}(\gamma, \gamma') < \varepsilon.$$

Furthermore, the length of the curve γ between the points $\gamma_\tau^\theta(t)$ and $\gamma_{\tau'}^{\theta'}(t)$—and hence the distance $\rho(\gamma_\tau^\theta(t), \gamma_{\tau'}^{\theta'}(t))$—does not exceed the larger of the following two arc lengths: (1) the arc length between the points $\gamma(\tau)$ and $\gamma(\tau')$ [which is equal to $|\tau - \tau'| J(\gamma) < \varepsilon J(\gamma)$]; and (2) the arc length between the points $\gamma(\theta)$ and $\gamma(\theta')$ [which is equal to $|\theta - \theta'| J(\gamma) < \varepsilon J(\gamma)$]. Therefore,

$$\rho\left(\gamma_\tau^\theta(t), \gamma_{\tau'}^{\theta'}(t)\right) < \varepsilon J(\gamma).$$

Since, on the other hand,

$$|J(\gamma_\tau^\theta) - J(\gamma_{\tau'}^{\theta'})| = |(\theta - \tau) J(\gamma) - (\theta' - \tau') J(\gamma)| < 2\varepsilon J(\gamma),$$

it follows that

$$\hat{\rho}(\gamma_\tau^\theta, \gamma_{\tau'}^{\theta'}) < 3\varepsilon J(\gamma).$$

Thus,

$$\hat{\rho}(\gamma_\tau^\theta, \gamma'^{\theta'}_{\tau'}) \leqslant \hat{\rho}(\gamma_\tau^\theta, \gamma_{\tau'}^{\theta'}) + \hat{\rho}(\gamma_{\tau'}^{\theta'}, \gamma'^{\theta'}_{\tau'}) < (1 + 3J(\gamma))\varepsilon.$$

Since $(1 + 3J(\gamma))\varepsilon \to 0$ as $\varepsilon \to 0$, this proves that the curve γ_θ^τ depends continuously on the curve γ and the numbers τ and θ.

This also proves that the deformation D_τ is continuous.

Now, consider the mapping

$$\alpha \circ \beta : [\varphi \leqslant b] \to [\varphi \leqslant b].$$

This mapping maps every point

$$p_* = (p_1, \ldots, p_s) \in [\varphi \leqslant b]$$

into the point

$$\left(\gamma\left(\frac{1}{s+1}\right), \ldots, \gamma\left(\frac{s}{s+1}\right)\right) \in [\varphi \leqslant b],$$

where $\gamma = \beta(p_*)$ is the broken geodesic

$$\gamma_{\overline{p}, p_1} * \gamma_{p_1, p_2} * \ldots * \gamma_{p_s, \overline{q}}$$

with vertices $p_1, \ldots, p_s$.

To show that this mapping is homotopic to the identity mapping $1 = 1_{[\varphi \leqslant b]}$ of the space $[\varphi \leqslant b]$, we must construct a continuous family (deformation) of continuous mappings

$$\Delta_\tau : [\varphi \leqslant b] \to [\varphi \leqslant b], \quad 0 \leqslant \tau \leqslant 1,$$

such that

$$\Delta_0 = 1, \quad \Delta_1 = \alpha \circ \beta.$$

Suppose that

$$0 < t_1 < \ldots < t_s < 1$$

are the values of the parameter t on the curve γ corresponding to the points $p_1, \ldots, p_s$:

$$p_1 = \gamma(t_1), \ldots, p_s = \gamma(t_s).$$

For arbitrary $\tau \in [0, 1]$, let us consider, in the manifold M_*, the point

$$\Delta_\tau(p_*) = \left(\gamma\left(t_1(1-\tau) + \frac{1}{s+1}\tau\right), \ldots \right.$$
$$\left. \ldots, \gamma\left(t_i(1-\tau) + \frac{i}{s+1}\tau\right), \ldots, \gamma\left(t_s(1-\tau) + \frac{s}{s+1}\tau\right)\right),$$

where $\gamma = \beta(p_*)$. Obviously,

The point $\Delta_\tau(p_*)$ depends continuously on the point p_* and the number τ.

Furthermore,

$$\Delta_0(p_*) = p_*, \quad \Delta_1(p_*) = (\alpha \circ \beta)(p_*).$$

Thus, to show that Δ_τ is the required deformation, it suffices to show that

$$\Delta_\tau : [\varphi \leqslant b] \to [\varphi \leqslant b],$$

that is, that

$\Delta_\tau(p_*) \in [\varphi \leqslant b]$ *for an arbitrary point* $p_* \in [\varphi \leqslant b]$.

Define

$$p_i^{(\tau)} = \gamma\left(t_i(1-\tau) + \frac{i}{s+1}\tau\right), \quad i = 1, \ldots, s.$$

Also, define

$$p_0^{(\tau)} = \bar{p}, \quad p_{s+1}^{(\tau)} = \bar{q} \text{ for all } \tau.$$

Obviously, for arbitrary $i = 0, 1, \ldots, s$, the distance

$$\rho\left(p_i^{(\tau)}, p_{i+1}^{(\tau)}\right)$$

between the points $p_i^{(\tau)}$ and $p_{i+1}^{(\tau)}$ does not exceed the length of the curve $\gamma = \beta(p_*)$ between these two points; that is, it does not exceed the number

$$J(\gamma)\left[t_{i+1}(1-\tau) + \frac{i+1}{s+1}\tau - t_i(1-\tau) - \frac{i}{s+1}\tau\right] =$$
$$= J(\gamma)(t_{i+1} - t_i)(1-\tau) + \frac{J(\gamma)}{s+1}\tau = \rho(p_i, p_{i+1})(1-\tau) + \frac{J(\gamma)}{s+1}\tau.$$

(We recall that the curve γ is the minimizing geodesic between the points p_i and p_{i+1}.) Consequently,

$$\begin{aligned}\varphi(\Delta_\tau(p_*)) = \sum_{i=0}^{s}\rho^2(p_i^{(\tau)}, p_{i+1}^{(\tau)}) \leqslant \\ \leqslant \sum_{i=0}^{s}\left[\rho(p_i, p_{i+1})(1-\tau)+\frac{J(\gamma)}{s+1}\tau\right]^2 = \\ =(1-\tau)^2\sum_{i=0}^{s}\rho^2(p_i, p_{i+1}) + \\ +2\frac{(1-\tau)\tau J(\gamma)}{s+1}\sum_{i=0}^{s}\rho(p_i, p_{i+1})+\frac{J(\gamma)^2}{(s+1)^2}\tau^2 = \\ =(1-\tau)^2\varphi(p_*)+2\frac{(1-\tau)\tau J(\gamma)^2}{s+1}+\frac{J(\gamma)^2\tau^2}{(s+1)^2} = \\ =\varphi(p_*)-2\tau\left(\varphi(p_*)-\frac{J(\gamma)^2}{s+1}\right)+\tau^2\left(\varphi(p^*)-\frac{2J(\gamma)^2}{s+1}+\frac{J(\gamma)^2}{(s+1)^2}\right) = \\ =\varphi(p_*)-2\tau\left(\varphi(p_*)-\frac{J(\gamma)^2}{s+1}\right)+ \\ +\tau^2\left(\varphi(p^*)-\frac{J(\gamma)^2}{s+1}\right)-\frac{\tau^2 J(\gamma)^2}{s+1}\left(1-\frac{1}{s+1}\right) \leqslant \\ \leqslant \varphi(p_*)-\tau(2-\tau)\left(\varphi(p_*)-\frac{J(\gamma)^2}{s+1}\right).\end{aligned}$$

But, in accordance with the Cauchy-Schwarz-Bunyakovskiy inequality (cf. footnote in section 2 of Chapter 5),

$$\varphi(p_*)-\frac{J(\gamma)^2}{s+1}=\sum_{i=0}^{s}\rho^2(p_i, p_{i+1})-\frac{1}{s+1}\left(\sum_{i=0}^{s}\rho(p_i, p_{i+1})\right)^2 \geqslant 0.$$

Therefore,

$$\varphi(\Delta_\tau(p_*)) \leqslant \varphi(p_*) \text{ for arbitrary } \tau \in [0, 1].$$

Thus, if $p_* \in [\varphi \leqslant b]$, i.e., if $\varphi(p_*) \leqslant b$, then $\varphi(\Delta_\tau(p_*)) \leqslant b$; that is, $\Delta_\tau(p_*) \in [\varphi \leqslant b]$. This proves that

The mappings α *and* β *are inverse homotopic equivalences of each other.*

Consequently, the reduction theorem is completely proven.

5. A generalized reduction theorem

Let N denote an arbitrary compact submanifold of the space M and suppose that N does not contain a given point $\bar{p} \in M$. Let us consider in the space Φ the subspace $\hat{\Omega}(N)$ consisting of all curves $\gamma \in \Phi$ such that

$$\gamma(0)=\bar{p}, \quad \gamma(1) \in N.$$

For arbitrary a, we denote by $[J \leqslant a]_N$ the subspace of the space $\hat{\Omega}(N)$ consisting of all curves $\gamma \in \hat{\Omega}(N)$ such that $J(\gamma) \leqslant a$.

For every geodesic $\gamma \in \hat{\Omega}(N)$ that is orthogonal for $t = 1$ to the submanifold N, we shall denote the indices of the point $t = 0$ and of the interval (0, 1) with respect to the submanifold N (cf. section 2 of the Appendix) by $\lambda_N^0(\gamma)$ and $\lambda_N^{0,1}(\gamma)$ respectively. We shall call the number $\lambda_N^0(\gamma)$ the corank of the geodesic γ and we shall call the number $\lambda_N^{0,1}(\gamma)$ its index (with respect to the submanifold N).

In this section, we shall prove the following theorem:

For an arbitrary number $a \geqslant \rho(\bar{p}, N)$, there exist a smooth manifold M_*, a smooth function φ defined on M_*, and a number b such that the space $[J \leqslant a]_N$ is of the same homotopic type as the space $[\varphi \leqslant b]$.

Furthermore, there exists a homotopic equivalence

$$\alpha : [J \leqslant a]_N \to [\varphi \leqslant b],$$

such that the curve $\gamma \in [J \leqslant a]_N$ is a geodesic orthogonal for $t = 1$ to the submanifold N (that is, $\dot{\gamma}(1) \perp N_{\gamma(1)}$) if and only if $\alpha(\gamma) \in [\varphi \leqslant b]$ is a critical point of the function φ. Finally,

(1) *The corank $c(\alpha(\gamma))$ of the point $\alpha(\gamma)$ is equal to the corank $\lambda_N^0(\gamma)$ of the geodesic γ with respect to the submanifold N;*

(2) *the index $h(\alpha(\gamma))$ of the point $\alpha(\gamma)$ is equal to the index $\lambda_N^{0,1}(\gamma)$ of the geodesic γ with respect to the submanifold N.*

This reduction theorem is analogous to the reduction theorem in section 1 and it reduces to that theorem when the submanifold N consists of a single point. Its proof is completely analogous to the proof of the theorem in section 1. Therefore, we shall not carry out the proof in detail but shall only note those places at which modifications are necessary.

First of all, for the manifold M_*, we must now take the Cartesian product

$$\underbrace{M \times M \times \ldots \times M}_{s \text{ times}} \times N$$

where s is the same number as in section 2. We define the number b just as in section 2, that is, by

$$b = \frac{a^2}{s+1},$$

and the function φ by

$$\varphi(p_*) = \rho^2(\bar{p}, p_1) + \ldots + \rho^2(p_i, p_{i+1}) + \ldots + \rho^2(p_s, q),$$
$$p_* = (p_1, \ldots, p_s, q) \in M_*, \quad p_1, \ldots, p_s \in M, \ q \in N.$$

Furthermore, we set

$$\alpha(\gamma) = \left(\gamma\left(\frac{1}{s+1}\right), \ldots, \gamma\left(\frac{s}{s+1}\right), \gamma^{(1)}\right)$$

and

$$\beta(\gamma) = \gamma_{\bar{p}, p_1} * \ldots * \gamma_{p_i, p_{i+1}} * \ldots * \gamma_{p_s, q},$$

where $\gamma_{p_i, p_{i+1}}$, for $i = 0, 1, \ldots, s$, is the minimizing geodesic connecting the points p_i and p_{i+1} (we assume that $p_0 = \bar{p}$ and $p_{s+1} = q$).

It is easy to show that all the results of section 2 remain valid (except that in the proof of assertions (1) and (2) of the reduction theorem we need to replace the reference to the results of Chapter 4 by the results of the appendix to Chapter 4). In formula (2) of section 2 appears the extra term

$$2\rho(p_s, q)(C_{s+1}, \gamma_s^+)$$

with $C_{s+1} \in N_q$, the vanishing of which is ensured by the orthogonality of the geodesic γ to the submanifold N at the point $t = 1$.

Finally, since the reasoning of sections 3 and 4 are of a general nature, they can also be applied without modification except that in the formula defining the deformation Δ_τ, we need to make an obvious modification.

Thus, the generalized reduction theorem is also valid.

Remark: An analogous reduction theorem holds for the more general space $\hat{\Omega}(N_0, N_1)$ consisting of curves beginning on one submanifold N_0 and ending on another submanifold N_1 (disjoint from the first). We leave as a rather lengthy exercise for the reader the details of the proof.

6. Comparison of the space $\hat{\Omega}$ with the space Ω

The reduction theorem establishes the connection between the variational properties of the geodesics in a compact Riemannian space M and the topological (more precisely, the homotopic) properties of the space $\hat{\Omega}$. However, a direct topological study of the space $\hat{\Omega}$ is made considerably more difficult because of the topologically noninvariant definition of this space. Therefore, it is not the space $\hat{\Omega}$ itself that is usually studied in topology but some larger space Ω consisting of all *continuous* curves $\gamma: [0,1] \to M$ connecting the points $\bar{p}$ and $\bar{q}$. Here, the space Ω is equipped with the so-called compact-open topology. In this topology, the open sets are the unions of finite intersections of sets of the form $W(K, U)$ consisting by definition of all curves $\gamma \in \Omega$ such that, for arbitrary t belonging to a given compact subset $K \subset [0,1]$, the point $\gamma(t)$ belongs to the given open set $U \subset M$. The chief purpose of the present section is to prove the following proposition, which reduces study of the space $\hat{\Omega}$ to the study of the space Ω:

The spaces $\hat{\Omega}$ and Ω are of the same homotopic type.

To prove this, let us first prove the following assertion:

The formula

$$\rho^*(\gamma_1, \gamma_2) = \max_{0 \leqslant t \leqslant 1} \rho(\gamma_1(t), \gamma_2(t)) \tag{1}$$

defines on the space Ω *a metric that induces a topology coincident with the compact-open topology of the space* Ω.

That formula (1) defines a metric on Ω is obvious (cf. the corresponding remarks in section 1 regarding the analogous metric $\hat{\rho}$ in the space $\hat{\Omega}$).

Let γ_0 denote an arbitrary curve in Ω and let V denote an arbitrary neighborhood of the curve γ_0 in the compact-open topology of the space Ω. Then, by definition, there exist compact sets $K_1, \ldots, K_s \subset [0,1]$ and open sets $U_1, \ldots, U_s \subset M$ such that

$$\gamma_0 \in W(K_1, U_1) \cap \ldots \cap W(K_s, U_s) \subset V.$$

For arbitrary $i = 1, \ldots, s$, the set $\gamma_0(K_i)$ is compact and contained in the open set U_i. Let δ denote a positive number such that the δ-neighborhood of the set $\gamma_0(K_i)$ is contained in the set U_i for every $i = 1, \ldots, s$. Then, it is clear that an arbitrary curve $\gamma \in \Omega$ such that $\rho^*(\gamma, \gamma_0) < \delta$ belongs to each of the sets $W(K_i, U_i)$ and hence belongs to their intersection. In other words, the δ-neighborhood $V_\delta(\gamma_0)$ of the curve γ_0 in the metric ρ^* is contained in the intersection of the sets $W(K_i, U_i)$. Consequently,

$$V_\delta(\gamma_0) \subset V. \tag{2}$$

Conversely, let δ be an arbitrary positive number. Since the curve γ_0 is continuous and the interval $[0,1]$ is compact, there exists a number δ_1 such that, for arbitrary points $t, \tau \in [0,1]$, the inequality $|t - \tau| < \delta_1$ implies the inequality $\rho(\gamma_0(t), \gamma_0(\tau)) < \delta/2$. Let $K_1, \ldots, K_s$ denote an arbitrary finite sequence of closed intervals of length less than δ_1 that cover the interval $[0, 1]$, and let $U_1, \ldots, U_s$ denote $(\delta/2)$-neighborhoods of the sets $\gamma_0(K_1), \ldots, \gamma_0(K_s)$ respectively. In the space Ω, let us consider the set

$$V = W(K_1, U_1) \cap \ldots \cap W(K_s, U_s).$$

By definition, this set is open in the compact-open topology of the space Ω. Let $\gamma \in V$. By hypothesis, for arbitrary $t \in [0,1]$, there exists an i such that $t \in K_i$. Since $\gamma \in V$, we have $\gamma \in W(K_i, U_i)$ and hence $\gamma(t) \in U_i$. Consequently, there exists a $\tau \in K_i$, such that $\rho(\gamma(t), \gamma_0(\tau)) < \delta/2$. On the other hand, since $|t - \tau| < \delta_1$, we have $\rho(\gamma_0(\tau), \gamma_0(t)) < \delta/2$. Therefore, $\rho(\gamma(t), \gamma_0(t)) < \delta$. Since t is arbitrary, this proves $\rho^*(\gamma, \gamma_0) < \delta$, that is, that the curve γ belongs to the δ-neighborhood $V_\delta(\gamma_0)$ of the curve γ_0 in the metric ρ^*. Consequently,

$$V \subset V_\delta(\gamma_0). \tag{3}$$

Relations (2) and (3) together mean that the topology induced by the metric ρ^* coincides with the compact-open topology. This completes the proof of the assertion made above.

Obviously, the space $\hat{\Omega}$ is contained in the space Ω. Let

$$i : \hat{\Omega} \to \Omega$$

denote the corresponding embedding mapping. Since the metric $\hat{\rho}$ defined in section 1 on the space $\hat{\Omega}$ is bounded from below by the metric ρ^* considered only on $\hat{\Omega}$), it follows from the assertion just proven that

The embedding mapping $i:\hat{\Omega}\to\Omega$ *is continuous.*

Let us now define on the space Ω a continuous function χ_τ that depends continuously on the parameter $\tau\in[0,1]$ by setting

$$\chi_\tau(\gamma)=\max_{|t_1-t_2|\leqslant\tau}\rho(\gamma(t_1),\gamma(t_2))-(1-\tau)d,\quad \gamma\in\Omega,$$

where $d=d_M$ is the number introduced in section 2. Obviously, for every curve $\gamma\in\Omega$, the function $\chi_\gamma(\tau)=\chi_\tau(\gamma)$ is strictly monotonic and

$$\chi_\gamma(0)<0\leqslant\chi_\gamma(1).$$

Therefore, for an arbitrary curve $\gamma\in\Omega$, there exists a (unique) number $\tau(\gamma)\in[0,1]$ such that

$$\chi_{\tau(\gamma)}(\gamma)=0.$$

It is easy to see that the function $\tau(\gamma)$ constructed in this way is a continuous function of the curve $\gamma\in\Omega$. It enjoys the following property:

For arbitrary points $t_1, t_2\in[0,1]$ *that satisfy the inequality*

$$|t_1-t_2|\leqslant\tau(\gamma),$$

the points $\gamma(t_1)$ *and* $\gamma(t_2)$ *in the space M can be connected by a unique minimizing curve.*

Proof: We have

$$\rho(\gamma(t_1),\gamma(t_2))\leqslant\max_{|t_1-t_2|\leqslant\tau(\gamma)}\rho(\gamma(t_1),\gamma(t_2))=(1-\tau(\gamma))d\leqslant d.$$

For every curve $\gamma\in\Omega$, consider the points in the interval [0, 1]

$$t_1=\frac{\tau(\gamma)}{2},\ t_2=2\frac{\tau(\gamma)}{2},\ \ldots,\ t_i=i\frac{\tau(\gamma)}{2},\ \ldots,\ t_s=s\frac{\tau(\gamma)}{2},$$

where s is the greatest integer such that $s\frac{\tau(\gamma)}{2}\leqslant 1$. Let $j(\gamma)$ denote a broken curve, consisting of minimizing geodesics, that connects the points $\bar{p}$ and $\bar{q}$, that has vertices at the points

$$\gamma(t_1),\ \gamma(t_2),\ \ldots,\ \gamma(t_i),\ \ldots,\ \gamma(t_s)$$

and that is parameterized (proportionately to arc length) in such a way that $j(\gamma)(0)=\bar{p}$ and $j(\gamma)(1)=\bar{q}$. According to the assertion just proven, the broken curve $j(\gamma)$ is uniquely determined by the curve γ. Since $j(\gamma)\in\Omega$, we have defined a mapping

$$j:\Omega\to\hat{\Omega}.$$

One can easily show that

The mapping $j:\Omega\to\hat{\Omega}$ is continuous.

(Cf. the corresponding assertion in section 3 for the analytic mapping $\beta\circ\alpha$.)

Finally, it is easy to see that

The mappings ι and j are mutually inverse homotopic equivalences,

The corresponding deformations D_τ and Δ_τ are completely analogous to the deformations D_τ considered in section 4. We leave the detailed construction of them to the reader.

Remark: Coincidence of homotopic type of the spaces $\hat{\Omega}$ and Ω also obtains for *noncompact* Riemannian spaces M. To prove this, one need only prove the existence of the function $\tau(\gamma)$ enjoying the property indicated above, which ensures single-valuedness of the mapping j. For this, in turn, it is sufficient to prove the existence of the function $\chi_\tau(\gamma)$. To do this, let us consider, on the space M, an arbitrary continuous positive function f such that all the spaces $[f\leqslant a]$ are compact. Obviously, for arbitrary positive a, there exists a positive number $d(a)$ such that any two points $p,\ q\in[f\leqslant a]$ such that $\rho(p,q)<d(a)$ can be connected in M by a unique minimizing curve. Obviously, the function $d(a)$ is a monotonic positive function of a. Therefore, there exists a continuous function $d_1(a)$ such that

$$0<d_1(a)\leqslant d(a)$$

for arbitrary $a\geqslant 0$. Now, let us define the function $\chi_\tau(\gamma)$, where $\gamma\in\Omega$, by

$$\chi_\tau(\gamma)=\max_{|t_1-t_2|\leqslant\tau}\rho(\gamma(t_1),\ \gamma(t_2))-(1-\tau)\,d_1\Big(\max_{0\leqslant t\leqslant 1}f(\gamma(t))\Big).$$

Obviously, this function possesses all the properties necessary to ensure the existence of the function $\tau(\gamma)$.

INDEX

A CATALOGUE OF SELECTED DOVER BOOKS IN ALL FIELDS OF INTEREST

A CATALOGUE OF SELECTED DOVER BOOKS IN ALL FIELDS OF INTEREST

THE EARLY WORK OF AUBREY BEARDSLEY, Aubrey Beardsley. 157 plates, 2 in color: *Manon Lescaut, Madame Bovary, Morte Darthur, Salome,* other. Introduction by H. Marillier. 182pp. 8⅛ x 11. 21816-3 Pa. $4.50

THE LATER WORK OF AUBREY BEARDSLEY, Aubrey Beardsley. Exotic masterpieces of full maturity: *Venus and Tannhauser, Lysistrata, Rape of the Lock, Volpone,* Savoy material, etc. 174 plates, 2 in color. 186pp. 8⅛ x 11. 21817-1 Pa. $5.95

THOMAS NAST'S CHRISTMAS DRAWINGS, Thomas Nast. Almost all Christmas drawings by creator of image of Santa Claus as we know it, and one of America's foremost illustrators and political cartoonists. 66 illustrations. 3 illustrations in color on covers. 96pp. 8⅜ x 11¼.
23660-9 Pa. $3.50

THE DORÉ ILLUSTRATIONS FOR DANTE'S DIVINE COMEDY, Gustave Doré. All 135 plates from Inferno, Purgatory, Paradise; fantastic tortures, infernal landscapes, celestial wonders. Each plate with appropriate (translated) verses. 141pp. 9 x 12. 23231-X Pa. $4.50

DORÉ'S ILLUSTRATIONS FOR RABELAIS, Gustave Doré. 252 striking illustrations of *Gargantua and Pantagruel* books by foremost 19th-century illustrator. Including 60 plates, 192 delightful smaller illustrations. 153pp. 9 x 12. 23656-0 Pa. $5.00

LONDON: A PILGRIMAGE, Gustave Doré, Blanchard Jerrold. Squalor, riches, misery, beauty of mid-Victorian metropolis; 55 wonderful plates, 125 other illustrations, full social, cultural text by Jerrold. 191pp. of text. 9⅜ x 12¼. 22306-X Pa. $7.00

THE RIME OF THE ANCIENT MARINER, Gustave Doré, S. T. Coleridge. Dore's finest work, 34 plates capture moods, subtleties of poem. Full text. Introduction by Millicent Rose. 77pp. 9¼ x 12. 22305-1 Pa. $3.50

THE DORE BIBLE ILLUSTRATIONS, Gustave Doré. All wonderful, detailed plates: Adam and Eve, Flood, Babylon, Life of Jesus, etc. Brief King James text with each plate. Introduction by Millicent Rose. 241 plates. 241pp. 9 x 12. 23004-X Pa. $6.00

THE COMPLETE ENGRAVINGS, ETCHINGS AND DRYPOINTS OF ALBRECHT DURER. "Knight, Death and Devil"; "Melencolia," and more—all Dürer's known works in all three media, including 6 works formerly attributed to him. 120 plates. 235pp. 8⅜ x 11¼.
22851-7 Pa. $6.50

MECHANICK EXERCISES ON THE WHOLE ART OF PRINTING, Joseph Moxon. First complete book (1683-4) ever written about typography, a compendium of everything known about printing at the latter part of 17th century. Reprint of 2nd (1962) Oxford Univ. Press edition. 74 illustrations. Total of 550pp. 6⅛ x 9¼. 23617-X Pa. $7.95

PRINCIPLES OF ORCHESTRATION, Nikolay Rimsky-Korsakov. Great classical orchestrator provides fundamentals of tonal resonance, progression of parts, voice and orchestra, tutti effects, much else in major document. 330pp. of musical excerpts. 489pp. 6½ x 9¼. 21266-1 Pa. **$7.50**

TRISTAN UND ISOLDE, Richard Wagner. Full orchestral score with complete instrumentation. Do not confuse with piano reduction. Commentary by Felix Mottl, great Wagnerian conductor and scholar. Study score. 655pp. 8⅛ x 11. 22915-7 Pa. $13.95

REQUIEM IN FULL SCORE, Giuseppe Verdi. Immensely popular with choral groups and music lovers. Republication of edition published by C. F. Peters, Leipzig, n. d. German frontmaker in English translation. Glossary. Text in Latin. Study score. 204pp. 9⅜ x 12¼.
23682-X Pa. $6.00

COMPLETE CHAMBER MUSIC FOR STRINGS, Felix Mendelssohn. All of Mendelssohn's chamber music: Octet, 2 Quintets, 6 Quartets, and Four Pieces for String Quartet. (Nothing with piano is included). Complete works edition (1874-7). Study score. 283 pp. 9⅜ x 12¼.
23679-X Pa. **$7.50**

POPULAR SONGS OF NINETEENTH-CENTURY AMERICA, edited by Richard Jackson. 64 most important songs: "Old Oaken Bucket," "Arkansas Traveler," "Yellow Rose of Texas," etc. Authentic original sheet music, full introduction and commentaries. 290pp. 9 x 12. 23270-0 Pa. **$7.95**

COLLECTED PIANO WORKS, Scott Joplin. Edited by Vera Brodsky Lawrence. Practically all of Joplin's piano works—rags, two-steps, marches, waltzes, etc., 51 works in all. Extensive introduction by Rudi Blesh. Total of 345pp. 9 x 12. 23106-2 Pa. $14.95

BASIC PRINCIPLES OF CLASSICAL BALLET, Agrippina Vaganova. Great Russian theoretician, teacher explains methods for teaching classical ballet; incorporates best from French, Italian, Russian schools. 118 illustrations. 175pp. 5⅜ x 8½. 22036-2 Pa. $2.50

CHINESE CHARACTERS, L. Wieger. Rich analysis of 2300 characters according to traditional systems into primitives. Historical-semantic analysis to phonetics (Classical Mandarin) and radicals. 820pp. 6⅛ x 9¼.
21321-8 Pa. $10.00

EGYPTIAN LANGUAGE: EASY LESSONS IN EGYPTIAN HIEROGLYPHICS, E. A. Wallis Budge. Foremost Egyptologist offers Egyptian grammar, explanation of hieroglyphics, many reading texts, dictionary of symbols. 246pp. 5 x 7½. (Available in U.S. only)
21394-3 Clothbd. $7.50

AN ETYMOLOGICAL DICTIONARY OF MODERN ENGLISH, Ernest Weekley. Richest, fullest work, by foremost British lexicographer. Detailed word histories. Inexhaustible. Do not confuse this with *Concise Etymological Dictionary,* which is abridged. Total of 856pp. 6½ x 9¼.
21873-2, 21874-0 Pa., Two-vol. set $12.00